Synthesis Lectures on Computer Science

The series publishes short books on general computer science topics that will appeal to advanced students, researchers, and practitioners in a variety of areas within computer science.

Lihui Liu · Hanghang Tong

Neural Symbolic Knowledge Graph Reasoning

A Pathway Towards Neural Symbolic AI

Springer

Lihui Liu
Department of Computer Science
Wayne State University
Detroit, MI, USA

Hanghang Tong
Siebel School of Computing and Data Science
University of Illinois Urbana-Champaign
Urbana, IL, USA

ISSN 1932-1228 ISSN 1932-1686 (electronic)
Synthesis Lectures on Computer Science
ISBN 978-3-032-15857-4 ISBN 978-3-032-15858-1 (eBook)
https://doi.org/10.1007/978-3-032-15858-1

This Springer imprint is published by the registered company Springer Nature Switzerland AG
The registered company address is: Gewerbestrasse 11, 6330 Cham, Switzerland

I would like to express my deepest gratitude to my family for their unconditional love and encouragement, to my friends for their constant companionship and inspiration, and to my advisor and labmates for their invaluable support throughout my research journey.

Preface

Artificial Intelligence (AI) has been transforming the way we live, work, and interact with the world. Among the many emerging directions of AI, *Neural-Symbolic AI* has gained significant attention in recent years. By combining the representational power of deep learning with the logical rigor of symbolic reasoning, it promises the next generation of AI systems that are more explainable, trustworthy, and versatile. Such systems are expected to revolutionize a wide range of high-impact applications, including code generation, question answering, and drug discovery.

To fully unleash the potential of neural-symbolic reasoning, a fundamental challenge lies in how symbolic knowledge is represented and integrated with neural models. Among various approaches, *knowledge graphs* (KGs) have emerged as a powerful and versatile framework for organizing and connecting real-world information. Acknowledge graph represents a collection of facts in the form of triples, each consisting of a subject, predicate, and object. This structured representation captures semantic relationships among entities and concepts, enabling machines to perform reasoning and inference over interconnected knowledge. Knowledge graphs have been successfully applied in diverse areas such as search engines, recommender systems, and knowledge graph question answering (KGQA).

The primary goal of knowledge graph reasoning is to derive meaningful insights from the available data—discovering hidden patterns, explaining existing knowledge, or inferring new information from known facts. Reasoning over KGs using neural network models naturally embodies the spirit of neural-symbolic AI, where symbolic structures guide logical reasoning while neural networks contribute generalization and learning capabilities.

Despite remarkable progress in knowledge graph reasoning, several open challenges remain to be addressed. First, the background knowledge graph is often vast and incomplete, creating difficulties for efficient and accurate reasoning. Explicit knowledge is stored as triples, while implicit knowledge is embedded in graph structures, paths, or subgraphs. Moreover, as the scale of KGs continues to grow, developing efficient reasoning strategies remains a key research problem. Second, the inputs to various reasoning tasks can differ significantly—queries may be ambiguous, uncertain, or iterative—making the reasoning process inherently more complex. Third, relations in KGs exhibit diverse logical properties, such as transitivity, symmetry, and asymmetry. Designing models that can capture these relational characteristics while maintaining generalization across tasks and across knowledge graphs is an ongoing challenge. Addressing these issues is essential for the continued advancement of neural-symbolic reasoning on knowledge graphs.

This book systematically explores the problem of knowledge graph reasoning through the integration of neural and symbolic techniques. The discussion is organized around five key research questions that represent different reasoning scenarios under varying conditions of completeness and uncertainty. The first question investigates how to design an efficient reasoning model when the knowledge graph is complete and the query is accurate. This involves developing algorithms capable of traversing large KGs quickly while accurately identifying relevant entities and relationships. The second question considers the case where the knowledge graph is complete but the query itself is uncertain. Here, the objective is to detect inconsistencies between the query and the KG, ensuring the reliability of reasoning outcomes and identifying potential misinformation. The third question addresses the challenge of incomplete knowledge graphs when the query is accurate, focusing on methods to mitigate incompleteness by leveraging existing knowledge and employing neural-symbolic reasoning techniques to infer missing facts. The fourth question examines scenarios where the knowledge graph is incomplete and the query is ambiguous, requiring models that can integrate limited symbolic information with contextual clues from natural language to accurately interpret and answer the query. Finally, the fifth question explores dynamic reasoning environments in which both the knowledge graph and the queries evolve over time, calling for iterative reasoning strategies that adapt continuously while preserving accuracy and efficiency.

Several representative methods are presented throughout this book to illustrate how these questions can be addressed. For efficient reasoning over complete graphs, a symbolic subgraph matching approach has been developed to accelerate querying in largescale KGs. To detect inconsistencies and identify fake information, a symbolic graph kernel-based method is proposed. To alleviate incompleteness in knowledge graphs, two neural-symbolic reasoning approaches are introduced that exploit latent relational structures. For handling ambiguous queries, a hybrid neural-symbolic framework is designed to combine limited symbolic KG information with contextual cues from the query itself. In dynamic reasoning environments, a reinforcement learning-based conversational question-answering algorithm is developed to iteratively refine answers based on symbolic paths in the KG.

Together, these developments form a comprehensive framework for understanding and advancing neural-symbolic reasoning on knowledge graphs, contributing toward the long-term goal of building AI systems that are not only intelligent but also transparent, interpretable, and trustworthy.

Detroit, USA
October 2025

Lihui Liu

Acknowledgements First and foremost, I would like to express my deepest gratitude to Dr. Hanghang Tong for his guidance, invaluable advice, and unwavering support throughout the development of this book. His encouragement and patience allowed me to explore complex ideas freely, while his rigorous approach to research instilled in me the importance of clarity, precision, and intellectual honesty. Beyond his academic guidance, his gentle personality, generosity, and thoughtfulness have been a source of inspiration. The insights and perspectives I gained from countless discussions with him have not only shaped the content of this book but have also enriched my understanding of the broader scientific and intellectual landscape. This book would simply not have been possible without his mentorship and support, and his influence will continue to guide me in my future endeavors.

I am also profoundly grateful to the members of the iDEA and iSAIL Labs, whose camaraderie, collaboration, and encouragement have been invaluable throughout this journey. Jingrui He, Xing Su, Liangyue Li, Chen Chen, Si Zhang, Boxin Du, Qinghai Zhou, Jian Kang, Zhe Xu, Baoyu Jing, Yuchen Yan, Shengyu Feng, Yuheng Zhang, Derek Wang, Yian Wang, Zhichen Zeng, Ishika Agarwal, Eunice Chan, Xinyu He, Blaine Hill, Zhining Liu, Ruizhong Qiu, Hyunsik Yoo, Alex Zheng, David Zhou, Ruike Zhu, Xiao Lin, Yaojing Wang, Yu Wang, Ziye Zhu, Hao Wang, Haoran Li, Shweta Jain, Tianwen Chen, Yunyong Ko, Dawei Zhou, Yao Zhou, Arun Reddy Nelakurthi, Xu Liu, Jun Wu, Lecheng Zheng, Xue Hu, Dongqi Fu, Pei Yang, Yikun Ban, Yunzhe Qi, Haonan Wang, Ziwei Wu, Wenxuan Bao, Tianxin Wei, Xinrui He, Isaac Joy, and Zihao Li have made this journey intellectually stimulating and personally rewarding. Their willingness to share ideas, provide feedback, and collaborate on challenging problems created an environment that nurtured creativity and persistence. Beyond their academic support, I cherish the friendships, laughter, and countless memorable moments we shared. I would also like to extend my thanks to all my friends, both within and outside the University of Illinois Urbana-Champaign, who have been a source of joy, inspiration, and encouragement throughout this journey. Their companionship has made the process of writing this book more enjoyable and meaningful.

Finally, I am profoundly grateful to my beloved parents for their unwavering love, steadfast support, and constant belief in me. From the earliest days, they instilled in me values of curiosity, discipline, and perseverance that have guided every step of my intellectual journey. Their sacrifices, encouragement, and wisdom have been my foundation, shaping my character and my approach to challenges. They have been a constant source of strength, providing guidance when I faced difficulties and celebrating my achievements with genuine joy. I owe them an immeasurable debt of gratitude for nurturing my growth, inspiring me to aim high, and teaching me the importance of integrity, compassion, and resilience. Their influence extends far beyond this book, and I carry their lessons and values with me in every endeavor I undertake.

In addition to these individuals, I am grateful to the broader community of colleagues, collaborators, and mentors who have contributed ideas, insights, and support in various forms. Each conversation, suggestion, and shared resource has enriched this work and helped shape the ideas presented within these pages. Writing this book has been a collaborative journey in spirit, drawing upon the generosity and expertise of many people, and I am deeply thankful to all who have played a role, whether directly or indirectly, in making it possible.

Competing Interests The authors have no competing interests to declare that are relevant to the content of this manuscript.

Contents

Part I
Introduction

1 Introduction

1.1 Motivation

Artificial intelligence (AI) has been transforming the way we live, work, and interact with the world. Under the umbrella of AI, there is a special type called neural AI, which is based on neural networks. Neural AI is efficient but lacks interpretability. Another type of AI is called symbolic AI, which is based on high-level symbolic representation. It is interpretable but time-consuming. Recently, there is a new trend to combine these two approaches into a new category called neural symbolic AI. This new approach leverages the advantages of both neural AI and symbolic AI. Despite the promise of neural symbolic AI, one problem is how to represent real-world knowledge effectively. Knowledge graphs (KGs) are collections of real-world facts represented as nodes (representing real-world entities, events, and objects) and edges (denoting relationships between nodes). They are a good way to represent real-world knowledge. Thus, reasoning on knowledge graphs can naturally leverage the advantages of neural symbolic AI.

The use of knowledge graphs has gained significant traction in a wide variety of applications, ranging from recommender systems [39], question answering [91], to fact-checking [92]. By leveraging the wealth of information contained within knowledge graphs, it is possible to greatly enhance various downstream tasks, making reasoning over knowledge graphs an area of increasing interest. Knowledge graph reasoning refers to the process of inferring new knowledge and discovering relationships between entities in a knowledge graph [89, 90]. The research goal of neural symbolic knowledge graph reasoning is to enable machines to reason and draw symbolic logical conclusions based on the explicit and/or implicit information stored in the knowledge graph with the help of neural network models. Knowledge graph reasoning is a fundamental aspect of artificial intelligence and machine learning, as it enables these systems to learn from and reason with structured knowledge graph data [76].

L. Liu and H. Tong, *Neural Symbolic Knowledge Graph Reasoning*, Synthesis Lectures on Computer Science, https://doi.org/10.1007/978-3-032-15858-1_1

However, despite its popularity, there are several unique challenges associated with it, including the background knowledge graph, the reasoning input and the reasoning model. First (Background Knowledge Graph), although knowledge graphs have been extensively used in various applications, real-world knowledge graphs tend to be vast and incomplete. For instance, Google's Knowledge Graph contains millions of entities and hundreds of millions of relations between them. On top of the enormous size of this knowledge graph, it has numerous missing or incomplete triples, i.e., statements about entities and their relationships. This incompleteness can lead to limitations when using the knowledge graph to answer complex queries or make informed decisions based on the available information. Most knowledge graphs are incomplete, meaning that they may not capture all the relevant knowledge required for reasoning. As a result, reasoning on incomplete knowledge graphs can be difficult. Moreover, while most knowledge is explicitly stored as triplets in the knowledge graph, much useful implicit knowledge is hidden within the graph structure, paths, or subgraphs, making it difficult for the reasoning model to utilize it effectively. Additionally, real-world knowledge graphs often evolve over time, which presents an additional challenge. As new information becomes available, knowledge graphs need to be updated accordingly to reflect the latest state of the domain they represent. Second (Reasoning Input), in some KG reasoning applications, users may be unfamiliar with the background knowledge graph, leading to the possibility of asking ambiguous questions that can make KG reasoning tasks more challenging. According to the research [46], 80% of time, two people will use different vocabulary to speak the same thing. Moreover, some applications require iterative reasoning, where users ask several related questions in sequence, further increasing the complexity of the task. For example, traditional KGQA approaches often only consider single-shot questions [126], rather than the iterative nature of real-world conversation with a QA system. Conversational question answering (ConvQA) addresses this gap by allowing users to interact with the QA system through conversation, as seen in popular applications like Apple's Siri, Amazon's Alexa and OpenAI's ChatGPT. Third (Reasoning Model), due to the varied properties of relations in knowledge graphs, such as transitivity, symmetry, and asymmetry, designing an all-round neural symbolic KG reasoning model that fits all these properties can be challenging. Furthermore, the sheer volume of knowledge graphs presents significant computational hurdles for efficient reasoning. Moreover, the generalization ability of the reasoning model is critical for handling new, unseen data effectively, and the model's ability to solve multiple tasks is also a significant challenge in designing a robust neural symbolic knowledge graph reasoning system. Finally, determining if the model can transfer its knowledge to other knowledge graphs or domains is a crucial factor in assessing the model's usefulness and practicality.

1.2 Overview of Research

To address the above challenges, the goal of my research is to design novel neural symbolic algorithms which can support a variety of knowledge graph reasoning tasks. In particular, my research aims to understand and harness the input query in knowledge graph reasoning process with neural symbolic techniques. This is because, despite substantial progress, previous research on knowledge graph reasoning has primarily concentrated on reasoning for fixed and precise input queries [11], often overlooking the inherent ambiguity and dynamic nature of input queries [82, 87]. For instance, in the knowledge graph completion task, it is typically assumed that an exact partial triplet of the form $(h, r, ?)$ or $(h, ?, r)$ is provided, while question answering tasks presume the input query to be unambiguous and flawless. In many real-world scenarios, however, input queries may be ambiguous, vague, and dynamic, posing significant challenges to existing knowledge graph reasoning methodologies. By accurately understanding and modeling such ambiguous, noisy, and dynamic input queries, it will bring an exponential growth of the complexity, the scale, and the reasoning power to knowledge graph reasoning models and systems. More specifically, we explore the following five distinctive but internally correlated research problems.

- Q1: when the knowledge graph is complete and the query is accurate, how to design an efficient reasoning model?
- Q2: when the knowledge graph is complete and the query is uncertain, how to detect the inconsistency in the query?
- Q3: when the knowledge graph is incomplete and the query is accurate, how to mitigate the KG incompleteness?
- Q4: when the knowledge graph is incomplete and the query is ambiguous, how to find answer accurately?
- Q5: when the knowledge graph is incomplete and the query is dynamic, how to find answer iteratively?

To study the previous five questions, I have conducted the following works.

- To address Q1, this book proposes an index-based symbolic algorithm (G- FINDER) to find the top-k approximate matching subgraphs. At the heart of the proposed algorithm are two techniques, including (1) a novel auxiliary data structure (LOOKUP-TABLE) in conjunction with a neighborhood expansion method to effectively and efficiently index candidate vertices, and (2) a dynamic filtering and refinement strategy to prune the false candidates at an early stage.
- To address Q2, this book proposes a symbolic graph kernel-based system to detect the inconsistency of the input query. The proposed method leverages a gradient-based influence function to measure the importance of each element in the query, predicting the

inconsistency based on these important elements and their corresponding information in the knowledge graph.

- To address Q3, this book proposes two neural symbolic reasoning methods. The first method uses a multi-task learning model to simultaneously perform knowledge graph completion and question answering, effectively handling queries on incomplete knowledge graphs. The second method combines large language models with knowledge graph reasoning to answer complex logic queries. Here, the information stored in the large language model is used to complete the knowledge graph, while the knowledge graph helps reduce the hallucination problem in large language models.
- To address Q4, we propose PREFNET, a neural symbolic approach designed to answer ambiguous queries. PREFNET uses variational Bayesian inference to understand the user's query intention. It then returns high-quality inferred queries to the users, making their search easier. Additionally, we propose a prompt-tuning method to match two ambiguous entities. This method uses both entity-level and attribute-level information to effectively solve the problem.
- To address Q5, we propose a neural symbolic reinforcement learning (RL) based model, CORNNET, which utilizes question reformulations generated by large language models (LLMs) to improve ConvQA performance. CORNNET adopts a teacher-student architecture where a teacher model learns question representations using human writing reformulations, and a student model to mimic the teacher model's output via reformulations generated by LLMs. The learned question representation is then used by a reinforcement learning model to locate the correct answer in a KG. Extensive experimental results show that CORNNET outperforms state-of-the-art ConvQA models.

1.3 Organization

The remainder of this book is organized as follows. In Chap. 2, we commence with a comprehensive review of the pertinent literature on neural symbolic knowledge graph reasoning methods, addressing key challenges from diverse perspectives. Subsequently, in Chap. 3, we delve into our works that are centered on addressing the question of "how to design an efficient reasoning model when the knowledge graph is complete and the query is accurate?" Moving forward, in Chap. 4, we explore how to tackle the question "how to detect inconsistency inside the query when the knowledge graph is complete?" Then, in Chap. 5 and Chap. 6, we present our efforts towards answering the question of "how to mitigate knowledge graph incompleteness when the knowledge graph is incomplete and the query remains accurate?" Next, in Chaps. 7 and 8, we introduce our work that aims to address the challenge of "how to find answers accurately when the knowledge graph is incomplete and the query is ambiguous?" Furthermore, in Chap. 9, we delve into our works that seek to answer the question of "how to find answers iteratively when the knowledge graph is incomplete and the query is dynamic?" Finally, in Chap. 10, we bring this book to a conclusion by summarizing our findings and discussing potential future research directions.

Literature Review 2

In this chapter, following the research overview, we review the related literature from the following aspects, including (1) symbolic knowledge graph reasoning, (2) neural knowledge graph reasoning, (3) neural symbolic knowledge graph reasoning and (4) LLMs with knowledge graph reasoning.

2.1 Symbolic Knowledge Graph Reasoning

Symbolic reasoning in knowledge graphs refers to the process of deriving logical conclusions and making inferences based on symbolic representations of entities, relationships, and rules within the graph structure. In this context, symbols represent entities or concepts, while relationships denote connections or associations between them. Symbolic reasoning involves applying logical rules and operations to manipulate these symbols, enabling the system to perform tasks such as deductive reasoning, semantic inference, and knowledge integration. By leveraging symbolic representations and logical reasoning, knowledge graphs can facilitate complex problem-solving, semantic understanding, and decision-making in various domains, ranging from natural language processing to artificial intelligence applications.

Most subgraph matching-based methods fall under symbolic reasoning, which aims to find answers in the knowledge graph by matching the structure of the query graph. Subgraph matching can be divided into exact subgraph matching and inexact subgraph matching. Depending on the specific problem to solve, the exact subgraph matching algorithms can be further divided into two sub-groups. The algorithms in the first sub-group aim to find subgraphs in a database with many data graphs. Representative algorithms include gIndex [158], FG-Index [27], Tree+delta [170] and so on. They typically use graph mining techniques to find small frequent subgraphs (e.g. path, tree) from the database, and then use a

L. Liu and H. Tong, *Neural Symbolic Knowledge Graph Reasoning*, Synthesis Lectures on Computer Science, https://doi.org/10.1007/978-3-032-15858-1_2

filter-and-refine strategy to prune false data graphs. In this way, the search space can often be greatly reduced. The algorithms in the second sub-group mainly focus on finding an exact subgraph from a very large data graph. Representative algorithms includes Ullmann [31], CFL [10] and so on. The main idea behind these algorithms is iteratively trying to map nodes one by one from a query graph to a data graph, and backtracing if the attempt fails.

Inexact subgraph matching focuses on finding approximate subgraphs on a large data graph. There have been extensive studies on this problem. To name a few, Tong et al. [144] propose the best-effort pattern matching, which aims to maintain the topology of the query. Tian et al. [142] propose an approximate subgraph matching tool TALE with efficient indexing. Khan et al. [64] propose a heuristic approach NeMa based on a new definition of matching cost metric. Pientar et al. [114] propose an algorithm called MAGE which is an improved version of G-Ray. Different from G-Ray, it supports graphs with both node and edge attributes. Zhang et al. propose an inexact subgraph matching algorithm SAPPER [168] which utilizes the hybrid neighborhood unit structures in the index. Tian et al. [141] propose an approximate graph matching algorithm SAGA which employs a flexible graph distance model to measure similarities between graphs. He et al. [56] propose an index based algorithm called Closure-Tree to support both subgraph queries and similarity queries.

Besides subgraph matching, symbolic reasoning has been used in many other applications, such as network modeling [100, 101], rule learning [97, 98], computational fact-checking [83], and so on. For fact-checking, given *a claim* in the form of a triple within the knowledge graph, it reasons whether the claim is authentic or falsified by symbolic techniques. For example, in [125], the authors focused on checking the truthfulness of a given triple/claim by first transforming the knowledge graph into a weighted directed graph, and then extracting a so-called knowledge stream based on a maximum flow algorithm. In [129], an alternative method was developed to detect fake claims by learning the discriminative paths of specific predicates. Unlike [125], this is a supervised reasoning method since it requires different training datasets for different predicates. If the predicate in the claim does not exist in the training data, which is likely to be the case for detecting falsified claims in emerging news, the algorithm becomes inapplicable. In [83, 88], the authors proposed a random walk graph kernel-based method to check the truthfulness of claims. They first used an influence function to measure the importance of each element in the knowledge segments, and then made predictions according to these important elements. In 2015, GL Ciampaglia et al. [30] showed that the complexities of human fact-checking can be approximated quite well by finding the shortest path between concept nodes under properly defined semantic proximity metrics on knowledge graphs. The authors evaluated tens of thousands of claims on knowledge graphs extracted from Wikipedia. Many research works follow this direction with different techniques. Lin et al. [78] introduced ontological patterns in fact-checking for semantic and topological constraints. These constraints are represented as subgraph patterns that are used for querying the knowledge graph. To establish the ground truth of contextualized claims, Tchechmedjiev et al. released a large, up-to-date, and queryable corpus of structured information about claims and related metadata for fact-checking research, named ClaimsKG [140].

2.2 Neural Knowledge Graph Reasoning

Neural reasoning in knowledge graphs involves using neural network models to reason within the embedding space. Unlike traditional symbolic reasoning, which relies on explicit rules, patterns, and logical operations, neural reasoning leverages deep learning techniques to learn implicit knowledge within the embedding space. This allows knowledge graphs to capture complex, non-linear dependencies between entities and infer higher-level knowledge from the graph's interconnected nodes. A typical task in neural knowledge graph reasoning is knowledge graph embedding. Given an input query in the form of a partial triplet $(h, t, ?)$, knowledge graph embedding aims to predict missing links or entities based on existing information in the knowledge graph. The main idea is to learn a low-dimensional vector for each entity and predicate in the embedding space and use these embedding vectors for reasoning tasks.

Many KG embedding methods have been developed. The basic idea of TransE [11] is to view the relation r as the transition from the head entity to the tail entity. Mathematically, it means that ideally, the tail entity t should be the summation of the head entity and the relation. Another method is DistMult [162]. Similar as TransE, DistMult also embed entities and relations into vectors in the real/encludience space. Different from TransE, DistMult views the relation r as the elementwise weights of the head entity h. Its score function is defined as the weighted sum over all elements of the head entity by the corresponding elements in the relation. So, in DistMult, the ideal tail entity should be $h\dot{r}$. Another method ComplEx [147] embeds entities and relations in Complex vector space. Each embedding now has a real part and an imaginary part. Given a point z which is $x + iy$ in the embedding space, its conjugate $\bar{Z}$ is $x - iy$. The scoring function used in complex is very similar to that of distmult. But We replace t by its conjugate we only taking the real part of the function. Different from dot product, in Complex [147] Hermitian dot product $< h, r, \bar{t} >$ is asymmetric, where $\bar{t}$ is the complex conjugate of t. Thus it naturally captures the anti-symmetry. Another method is called RotatE [138]. The key idea of rotatE is to solve the limitations in previous methods. Similar to complex, rotatE also represent head and tail entities and relation in complex vector space, Different from complex, all the relation embedding are modelled as rotation from the head entity h to the tail entity t. Compared with other methods, RotatE can support different relation properties, such as symmetry/antisymmetry, inversion, and composition. Some additional methods, like BoxE [2] and KG2E [53], use geometric boxes or Gaussian distributions to represent entities.

2.3 Neural Symbolic Knowledge Graph Reasoning

Neural symbolic reasoning represents a fusion of neural network-based approaches with symbolic reasoning techniques, aiming to leverage the strengths of both paradigms in handling complex reasoning tasks. In this framework, neural networks are used to learn representations of symbolic entities and relationships within a knowledge graph, capturing both their semantic meanings and structural dependencies. These learned representations are then combined with symbolic reasoning mechanisms to perform logical inference and reasoning tasks. By integrating neural and symbolic components, neural symbolic reasoning approaches strive to overcome the limitations of each individual approach. Neural networks offer the ability to learn from data and handle uncertainty, while symbolic reasoning provides formal logic-based reasoning and interpretability. This hybrid approach has shown promise in various domains, including natural language understanding, knowledge graph reasoning, and automated theorem proving, by enabling more robust and flexible reasoning capabilities. Neural symbolic knowledge graph reasoning can be applied to various applications, such as question answering, recommender systems, network alignment [159, 160] and so on. In this section, we primarily discuss works related to handling different query types during the reasoning process.

When the input query is a natural language question, this problem is also known as Knowledge Graph Question Answering, which has been studied extensively. Wang et al. [152] proposed a Bayesian probability model combined with random walks to find the most similar concepts for a given query entity. Yang et al. [163] found that due to the lack of insight about the background knowledge graph, it is often hard for a user to precisely formulate a query. They developed a user-friendly knowledge graph search engine to support query formation and transformation. Jayaram et al. [58] proposed a knowledge graph query system called GQBE. Unlike other graph query systems, GQBE focuses on entity tuple queries, which consist of a list of entity tuples. To answer natural language sentence, another strategy is to transform the question into a query graph and search for the answer according to the query graph. For example, in [165], Xi et al. propose a model that contains a candidate query graphs ranking component and a true query graph generation component. By iteratively updating these two components, both components' performance can be improved. The query graph is finally generated by the second component and can be used to search the KG. In [86], Liu et al. propose a multi-task model to tackle KGQA and KGC simultaneously. Other methods, such as [103, 126], directly learn an embedding from the natural language sentence and search for answers in the embedding space.

When the input query is a graph query, [117] models different operations in the query graph as different neural networks and transforms the query process into an entity search problem in the embedding space. Some other works try to use reinforcement learning to solve the question answering problem. For example, Zhang et al. [167] use a knowledge graph as the environment and propose a reinforcement learning-based agent model to navigate the KG in order to find answers to input questions. Similarly, in [35, 79, 157], authors use reinforcement learning models to find paths in the knowledge graph for answering input queries. Other studies, such as [3, 12, 105, 115, 172], integrated reinforcement learning with other methods to create more human-like answers.

When the query is ambiguous, question rewriting is usually utilized to solve this problem. Question rewriting, which aims to reformulate an input ambiguous question into a more salient representation, is a common approach to answer ambiguous questions. This can improve the accuracy of search engine results or make a question more understandable for a natural language processing (NLP) system. In [148], a unidirectional Transformer decoder is proposed to automatically rewrite a user's input question to improve the performance of a conversational question answering system. In [41], the authors propose a Seq2Seq model to rewrite the current question according to the conversational history and introduce a new dataset named CANARD. In [42], query rewriting rules are mined from a background KG, and a query rewriting operator is used to generate a new question.

When the query is dynamic, this problem is also known as conversational Question Answering (ConvQA). Conversational Question Answering (ConvQA) aims to understand and respond to questions in a conversational context. Various approaches have been used to develop ConvQA systems. For instance, in [13], the authors employed reinforcement learning to train an agent that reformulates input questions to aid the system's understanding. In [51], an encoder-decoder model is used to transform natural language questions into logical queries for finding answers. In [61], a transformer model is used to generate logical forms and graph attention is introduced to identify entities in the query context. In [3], the authors proposed a data-driven approach for building task-oriented dialogue systems, specifically on the Alexa platform, using both annotated and unannotated data to provide an easy way for developers to train and deploy dialogue models while allowing for the integration of new data sources and the incorporation of new tasks and domains. Other systems, such as Google's Lambda [1], Apple's Siri, and OpenAI's ChatGPT, are also pursuing this task. Reinforcement learning and machine learning techniques can be used to solve various tasks [17–22, 69, 113]. Recently, some other works try to use reinforcement learning to tackle multi-turn conversations. For example, in [38], the authors proposed a multi-turn dialogue agent which helps users search Knowledge Bases (KBs) without composing complicated queries. In [63], multiple agents are used to search the background knowledge graph according to the input question and conversation history. In [167], instead of using a random walk agent, an adaptive path generator is developed with several atomic operations to sequentially generate the relation paths until the agent reaches the target entity. Besides, a semantic policy network is presented with both character-level and sentence-level information to better guide the agent. Despite great successes have been achieved, the current iterative reasoning methods still face several limitations and open challenges. These include the difficulty in handling incomplete or noisy knowledge graphs, ensuring that the generated responses are coherent and relevant to the user's needs, and the need for continuous learning and adaptation to handle new and evolving knowledge. Addressing these challenges would greatly enhance the effectiveness and usability of iterative knowledge graph reasoning.

2.4 LLMs with Knowledge Graph Reasoning

The landscape of large language models has evolved significantly with the development of diverse and powerful models. Among these models, GPT-2 [115] and GPT-3 [12] have gained immense popularity for their remarkable text generation capabilities and the ability to perform a wide range of tasks. LLaMA [145] and LLaMA2 [146] are state-of-the-art LLMs capable of generating code, and natural language about code, from both code and natural language prompts. BARD [99] is trained with the latest information from the Internet, so it has more data to gather information in real time. These diverse large language models represent an exciting spectrum of approaches, offering researchers and practitioners a rich toolkit to address different NLP tasks.

Recent work on using language models for reasoning tasks are mostly based on prompt engineering. For example, in [67], Kojima et al. find that adding some simple information `'Let's think step by step.'` can greatly improve LLMs' performance on a variety of downstream tasks. In [154], Wei et al. aim to answer questions by breaking them down into intermediate steps. They prompt the language model with similar examples where an answer is formed step-wise before providing the requested answer. Based on this idea, Yao et al. [164] sample multiple thoughts at each intermediate step and make the model perform deliberate decision making by considering multiple different reasoning paths. Besta et al. [8] further generalize this idea and model the information generated by an LLM as an arbitrary graph, where units of information ('LLM thoughts') are vertices, and edges correspond to dependencies between these vertices. Some other methods, such as [32, 171], propose different methods for prompt engineering. Some research efforts aim to combine knowledge graphs with large language models (LLMs) to answer natural language questions. Examples include ROG [96] and TOG [137]. However, these approaches are limited to addressing simple or multi-hop questions. Other approaches utilize retrieval-augmented generation [74] methods to tackle complex natural language questions, such as GraphRAG [40] and REPLUG [131]. Nevertheless, these methods are not specifically designed to handle complex logical queries. The only work focused on answering complex logical queries is KARL [28].

Part II
Knowledge Graph Reasoning

3 Accurate Query Answering with Symbolic Reasoning Over Complete KGs

When the knowledge graph is complete and the query is accurate, the goal is to design an efficient reasoning model that can quickly and precisely extract relevant information from the graph to provide accurate answers. This involves developing algorithms that can effectively navigate the graph, identify relevant entities and relationships, and perform complex reasoning tasks to arrive at the most accurate response. In this chapter, we introduce our works on approximate symbolic subgraph matching when the knowledge graph is complete.

3.1 Introduction

(Approximate) Subgraph matching is a core primitive across a number of disciplines, ranging from data mining, databases, information retrieval, computer vision to natural language processing. In data mining, frequent patterns can be found by approximate subgraph isomorphism [4]; in databases and information retrieval, a query to search for a set of inter-correlated research papers can be treated as a subgraph matching problem [55]; in computer vision, the symbol and object recognition can be formulated as an approximate subgraph matching problem with a tolerance of node merging and splitting, due to object overlapping or segmentation error [95]; in natural language processing, words in a lexical resource can be viewed as nodes in a graph, their relations can be viewed as edges between nodes, and paraphrases can then be seen as matching graphs [106]. Generally speaking, the existing work can be grouped into two categories, including the exact matching methods and the inexact matching methods. Each of these two types of methods has its own pros and cons, which we elaborate below.

L. Liu and H. Tong, *Neural Symbolic Knowledge Graph Reasoning*, Synthesis Lectures on Computer Science, https://doi.org/10.1007/978-3-032-15858-1_3

On one hand, extensive research effort has been devoted to exact matching, e.g., [9, 10, 34, 119]. The key ideas behind these methods are either pruning false candidates at an early stage and/or building indexes for promising candidates to accelerate the search. Through some effective pruning conditions, the search space can often be greatly reduced. However, many real graphs are noisy and incomplete. Consequently, the observed data graph might not contain any exact matching subgraph at all. In this case (i.e., whether or not exact matching exists is unknown), existing index-based exact matching methods (e.g., [10]) might iterate over *all* possible candidates in the index, which has an exponential time complexity before returning nothing. Moreover, in progressive search [5, 16], the user might not know what exactly s/he is looking for. In this case, exact matching, even if it exists, might not be desirable.

On the other hand, existing inexact subgraph matching methods often rely on certain heuristics to identify important seed nodes, and then progressively expand to neighbors of seeds [141, 142, 144]. Usually, these methods can return an approximate matching subgraph in a relatively short time. However, the greedy strategy of expanding the partial matching subgraph step by step could easily be diverted to a sub-optimal search path. For example, in Fig. 3.1, assuming that the seed node is v_0. Many algorithms (e.g. [142, 144]) will choose v_1 as the next candidate node because v_1 can be perfectly mapped to u_1. After that, it will expand to v_4 and v_5, and finally, return the subgraph induced by (v_0, v_1 ,v_4, v_5, v_8). But obviously, (v_0, v_2 v_3 ,v_6, v_7, v_9, v_{10}, v_{11}) is a better result because it only contains one intermediate vertex v_2.

Due to the noise and incompleteness of the data graph, and the uncertainty of the query graph, exact matching subgraphs may not exist or may not be desirable. Therefore, in this chapter, we mainly focus on approximate subgraph matching although our proposed algorithm is also capable to find exact matching. In particular, we propose a new algorithm called G-FINDER. The main idea of G-FINDER is to improve the matching accuracy by building

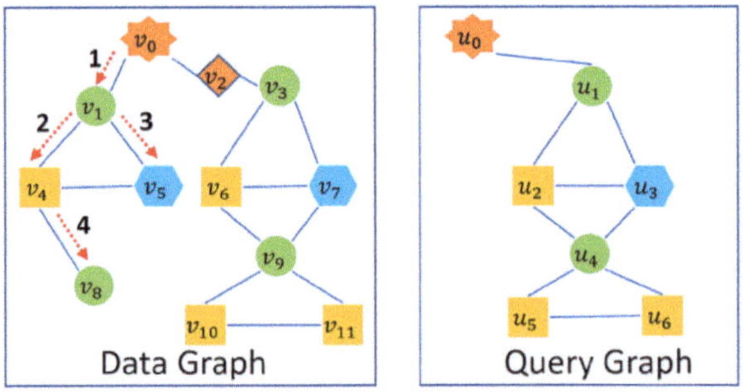

Fig. 3.1 An illustrative example. Left is the data graph and right is the query graph. Different node shapes and colors represent node attributes. Red dash arrows show the matching steps of TALE [142] which selects one important node at the first step. The proposed G-FINDER finds a matching subgraph induced by (v_0, v_2 v_3 ,v_6, v_7, v_9, v_{10}, v_{11})

index on data graph according to the query graph structure. Compared with existing inexact matching algorithms, the proposed G-FINDER embraces a novel auxiliary data structure called LOOKUP-TABLE (LTB), in conjunction with a neighborhood expansion method to index the candidate vertices of the query graph, and in the meanwhile it maintains the topology structure of the query graph. Furthermore, the proposed LTB can effectively prune false-positive candidates of query nodes for the purpose of computing an effective matching order. We conduct extensive experiments on various real datasets, demonstrating that G-FINDER achieves up to 30% better performance than the best baseline in F1-Score, and scales near-linearly with respect to the size of both the data graph and the query graph, particularly for data graphs with rich attribute information.

3.2 Problem Definition

Table 3.1 summarizes the main symbols and notations used in the chapter. For clarity, we refer to nodes in Q as nodes and that in $\mathcal{G}$ as vertices. Likewise, we call edges in Q as edges and that in $\mathcal{G}$ as links. We use u and $\hat{u}$ to refer to nodes of the query graph Q; and use v, $\hat{v}$ to refer to vertices of the data graph $\mathcal{G}$. A graph is denoted as a tuple $\{V, E, L\}$, where V is the node/vertex set, E is the edge/link set and L is an attribute function which maps a node/vertex or an edge/link to an attribute value. Given a query graph Q and a data graph $\mathcal{G}$, we say Q is the subgraph of $\mathcal{G}$ if there exists an exact subgraph matching from Q to $\mathcal{G}$ which satisfies the following condition.

Table 3.1 Notations and definition

Symbols	Definition
$Q = \{V_Q, E_Q, L_Q\}$	An attributed query graph
$\mathcal{G} = \{V_G, E_G, L_G\}$	An attributed data graph
$\mathcal{H}_i$	A partial matching with i nodes
Q_i	A partial query graph with i nodes
u, $\hat{u}$	Nodes in query graph
v, $\hat{v}$	Vertices in data graph
$\mathrm{m}(u_i) = v_i$	Matching function
N(u)	Neighborhood of query node u
C(u)	Candidate set of query node u
LTB(u)	LOOKUP-TABLE of query node u
LTBG	LOOKUP-TABLE-GRAPH formed by all LTBs
f(Q, $\mathcal{H}$)	The loss function

Definition 3.1 (*Exact Subgraph Matching* [158]). An exact subgraph matching (subgraph isomorphism) is an injective function $m()$: $V(Q) \longrightarrow V(\mathcal{G})$, which satisfies: (1) $\forall\, u \in V(Q)$, $m(u) \in V(\mathcal{G})$ and $L(u) = L(m(u))$; (2) $\forall\ (u_a, u_b) \in E(Q)$, $(m(u_a), m(u_b)) \in E(\mathcal{G})$ and $L(u_a, u_b) = L(m(u_a), m(u_b))$, where L is the attribute function.

Next, we introduce some basic concepts of approximate subgraph matching. Compared with exact subgraph matching, approximate subgraph matching can tolerate some missing query nodes, missing query edges and intermediate data vertices. We say a node u is a missing query node, if $u \in V(Q)$, but $m(u) \notin V(\mathcal{G})$. We say an edge (u_a, u_b) is a missing query edge, if $(u_a, u_b) \in E(Q)$, but $(m(u_a), m(u_b)) \notin E(\mathcal{G})$. We say a data vertex v_i returned by G-FINDER is an intermediate data vertex if it satisfies: (1) two nodes u_a, u_b in $V(Q)$ and $(u_a, u_b) \in E(Q)$, but $(m(u_a), m(u_b)) \notin E(\mathcal{G})$; (2) $(m(u_a), v_i) \in E(\mathcal{G})$ and $(v_i, m(u_b)) \in E(\mathcal{G})$. Last but not least, the remaining non-missing query nodes in Q and non-intermediate data vertices in the resulting subgraph returned by a matching algorithm which satisfy the mapping function $m()$ are referred to as matching pairs. For example, in Fig. 3.1, assuming the subgraph return by a matching algorithm is $(v_0, v_2, v_3, v_6, v_7, v_9, v_{10}, v_{11})$, then v_2 is an intermediate data vertex. If the matching subgraph is $(v_0, v_1, v_4, v_5, v_8)$, then u_5 and u_6 are missing query nodes, and (u_3, u_4), (u_4, u_5), (u_4, u_6), (u_5, u_6) are missing query edges, and (u_0, v_0) is a matching pair.

Another subtle difference between exact subgraph matching and approximate subgraph matching is the methodology to measure the goodness of a matching. For approximate matching, an intuitive way to measure the quality of a resulting matching subgraph $\mathcal{H}$ (e.g., found by G-FINDER) is to use a loss function to quantify the difference between the matching subgraph and the query graph Q. If the resulting matching subgraph has more intermediate data vertices, and the query graph Q has more missing query nodes and missing query edges, the loss function cost should be higher. In this chapter, we use a linear loss function defined as below.

Definition 3.2 (*Linear Loss Function*). Consider a query graph Q and a resulting matching subgraph $\mathcal{H}$. The linear loss function $f(Q, \mathcal{H})$ is defined as $f(Q, \mathcal{H}) = w_1 \times MN + w_2 \times ME + w_3 \times IN$, where MN is the number of missing query nodes in Q, ME is the number of missing query edges in Q, and IN is the number of intermediate data vertices in $\mathcal{H}$.

In the above loss function, w_1, w_2, w_3 are the weights for different types of inexact matching. By changing these weights, we can adjust the tolerance of the number of missing query nodes/edges and intermediate data vertices. For example, if w_1 is small, the results could tolerance more missing query nodes, and if $w_2 = \infty$, the results cannot tolerance any missing query edges. If an exact matching subgraph exists, the loss function cost should be 0.

In exact subgraph matching, a node u in Q can be mapped to a vertex v in $\mathcal{G}$ only if each of u's neighbors could be mapped to some of v's neighbors. However, this hard constraint can

Table 3.2 Comparison of subgraph matching methods

	Node attribute	Edge attribute	Exact matching	Missing nodes	Missing edges	Intermediate nodes	Index-based	Early pruning	Multi-graph
G-FINDER	✓	✓	✓	✓	✓	✓	✓	✓	✓
CFL [2]	✓	✗	✓	✗	✗	✗	✓	✓	✗
G-Ray [3]	✓	✗	✗	✓	✓	✓	✗	✗	✗
FIRST [4]	✓	✓	✗	✓	✓	✓	✗	✗	✗
MAGE [5]	✓	✓	✗	✓	✓	✓	✗	✗	✗
FilM [6]	✗	✗	✓	✗	✗	✗	✗	✓	✓
NeMa [7]	✓	✓	✓	✓	✓	✓	✓	✓	✗

Rows are methods and columns are desired properties

be relaxed in approximate subgraph matching. That is, a node u can be mapped to a vertex v even though their neighbors could not be perfectly matched with each other. For example, in Fig. 3.1, u_1 could be mapped to v_3, even though u_0 cannot match to any of v_3's neighbors. In the proposed G-FINDER, we allow a node in $\mathcal{Q}$ and a vertex in $\mathcal{G}$ to be mapped to each other if they satisfy the following conditions: (1) have the same attribute value, (2) bear a high node-vertex similarity, and (3) their neighbors also bear high node-vertex similarity, where the node-vertex similarity is defined as follows.

Definition 3.3 *Node-Vertex Similarity* is defined as $sim(v_i, u_i) = \frac{|(m(N(u_i))\cap N(v_i))|}{|N(u_i)|}$, where $N(v_i)$ denotes the neighbors of v_i, and $m(N(u_i)) \cap N(v_i)$ represents the nodes in $N(u_i)$ that are mapped to $N(v_i)$.

For example, in Fig. 3.1, since u_2 can be mapped to v_6, and u_3 can be mapped to v_7, the node-vertex similarity between v_3 and u_1 is $sim(v_3, u_1) = 2/3$.

Remarks. For both loss function and node-vertex similarity, there exist alternative choices. For example, there are a wealth of different graph similarity measures, e.g., graph edit distance [44], graph kernel [128] and embedding based graph similarity [110]. Likewise, there are also rich node similarity measures in the literature, e.g., matrix based method [5], graph neural networks based similarity [50]. In principle, the proposed G-FINDER algorithm can admit these alternative graph similarity as well as node similarity measures. In this chapter, we use these two relatively simple forms due to the efficiency consideration.

Finally, the *top-k* approximate subgraph matching problem can be formally defined as follows.

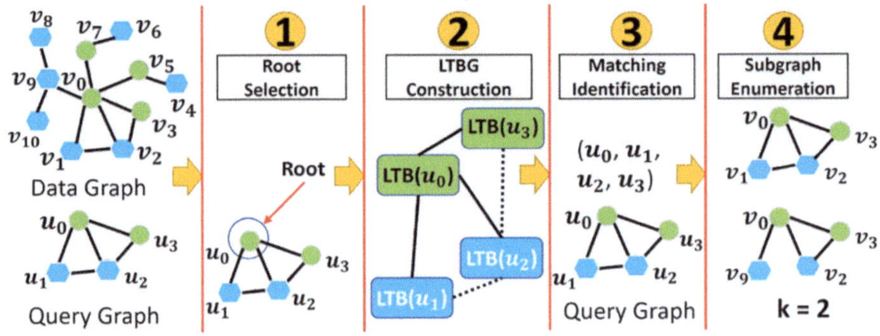

Fig. 3.2 The overall framework of G-FINDER

Problem 3.1 *top-k* Approximate Subgraph Matching:[1]

Given: (1) An attributed data graph $\mathcal{G}$, (2) an attributed query graph $\mathcal{Q}$, (3) the number of desired matching subgraphs k, and (4) a loss function $f()$;

Find: k approximate matching subgraphs that match the query graph $\mathcal{Q}$ as well as possible (have the smallest loss function cost).

3.3 Method Overview

The overall framework of G-FINDER contains four major steps which are illustrated in Fig. 3.2. Algorithm 1 presents the corresponding pseudo code, and Table 3.2 compares G-FINDER and existing algorithms, in terms of the input data graph, the ability for exact matching, the types of inexact matching, and computational efficiency.

First (Root Selection, Lines 3–4), G-FINDER selects a root node from the query graph. Second (LOOKUP-TABLE-GRAPH Construction, Line 7), a dynamic filtering and refinement strategy is used to construct LOOKUP-TABLE-GRAPH (short for LTBG) for the query graph. Third (Matching Identification, Line 8), a matching order is calculated based on the constructed LOOKUP-TABLE-GRAPH. Finally (Subgraph Enumeration, Line 9), G-FINDER searches LOOKUP-TABLE-GRAPH according to the matching order to return the *top-k* results which have the smallest loss function cost. In the remaining of this section, we highlight each of these four steps.

[1] The data graph and query graph can be multi-graph. For easier understanding, we mainly focus on graph with node attribute in this chapter.

Algorithm 1: G-FINDER

Input: a query $\mathcal{Q}$, a data graph $\mathcal{G}$, the number of desired matching subgraphs k and a loss function $f()$

Result: *top-k* approximate matching subgraphs

$(V_c, V_t) \leftarrow$ Core-Forest-Decompose($\mathcal{Q}$) // V_c is the core structure, and V_t is the forest structure. ;

(Query root u_r, Data root vertices S_{root}) $\leftarrow$ Root-Selection ;

Initialize top-k heap ;

for *each* $root_i \in S_{root}$ **do**

  LTBG $\leftarrow$ LTBG-Builder($\mathcal{Q}$, u_r, $root_i$, $\mathcal{G}$) ;

  Subgraph-Enumeration(k, LTBG, $\mathcal{Q}$, $\mathcal{G}$, M, $f()$) ;

Return *top-k* approximate matching subgraphs

3.3.1 Step 1: Root Selection

Given a query graph $\mathcal{Q}$, we first select a root node to start the matching process. For example, in Fig. 3.1, we can choose u_0 as the root. Many existing root selection methods can be used in the proposed G-FINDER algorithm. Basically, we would like to choose the root node which (1) has as few candidates as possible, and (2) is at the center of the query graph so that the diameter of the search space could be minimized [9]. With these two design objectives in mind, we first decompose the query graph into the core-forest structure.[2] After that, we choose the node $u_r \leftarrow \arg\min_u \frac{|C(u)|}{\deg(u)}$ from the core structure as the root, where $|C(u)|$ is the size of the candidate set of u in the data graph, and $\deg(u)$ is the degree of node u in the query graph.[3]

3.3.2 Step 2: LOOKUP-TABLE-GRAPH Construction

Following the root selection, the next step is to traverse the query graph to build index node by node. The key of this step is a novel data structure called LOOKUP-TABLE (short for LTB), which stores the information of the candidate vertices. Each node u in the query graph has a corresponding LTB(u), and all the LTBs will form a graph which shares the same topology as the query graph. We call this new graph as LOOKUP-TABLE-GRAPH (short for LTBG).

[2] The core structure of a graph $\mathcal{Q}$ is a maximum subgraph of $\mathcal{Q}$ where each node has at least 2 neighbors [10]. The remaining parts of the query graph is called the forest-structure of $\mathcal{Q}$. Figure 3.3a, b gives an example of the core-forst decomposition.

[3] For Fig. 3.1 in practice, G-FINDER will choose u_4 as the root node. However in order to explain the details of G-FINDER, we assume u_0 is the root node for all examples in this chapter. Moreover, the root node could not be a missing vertex.

For example, in column 3 of Fig. 3.2, LTB(u_0), LTB(u_1), LTB(u_2), and LTB(u_3) form a LOOKUP-TABLE-GRAPH for the input query.

The LTBG is built according to the traversing order of the query graph. Figure 3.3c, d gives an example of two traversal strategies, including BFS Tree and Dynamic-Tree. BFS is a common traversal strategy used by several existing exact subgraph matching methods [9, 10]. There are two kinds of edges in this strategy, including tree edges (TE) and non-tree edges (NTE). The edges that exist in both the BFS tree and the query graph are called tree edges. The edges that only exist in the query graph but not in the BFS tree are called non-tree edges. For example, in Fig. 3.3c, (u_2, u_4) is a tree edge, and (u_3, u_4) is a non-tree edge.

Although BFS is a common method for exact subgraph matching to traverse the query graph, it may not be suitable for approximate matching. For example, suppose we use BFS traversal strategy for the data graph in Fig. 3.1 and the query graph in Fig. 3.3a. Since u_7 has no candidates in the data graph, LTB(u_7) is empty, and consequently, both u_5 and u_6 will be ignored. Finally, the result graphs found by G-FINDER will be bad.

To tackle this issue for approximate subgraph matching, we propose a heap based dynamic traversal strategy to build LOOKUP-TABLE-GRAPH. We maintain a heap to store the current processing LTB and each time we pop the LTB with the smallest $\frac{|C(u)|}{\deg(u)}$ from the heap. We recursively build LOOKUP-TABLE for its un-visited neighbors and push them into the heap. We repeat this procedure until all nodes in the query graph are processed.

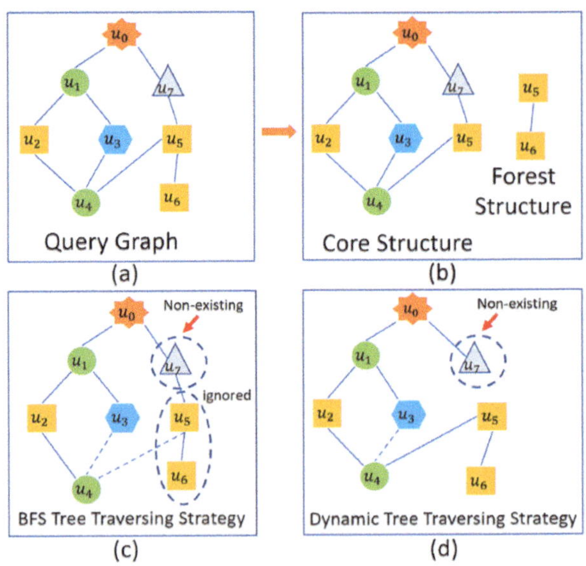

Fig. 3.3 The core-forest decomposition (**a**, **b**) and two traversal strategies (**c**, **d**). The dash lines represent non-tree edges

3.3.3 Step 3: Matching Identification

A matching order is the node sequence of the query graph, based on which we match the candidates in LTBG to the query nodes. For example, in Fig. 3.2, the matching order is (u_0, u_1, u_2, u_3). G-FINDER will first choose a good candidate v_0 in LTB(u_0), and then choose a good candidate v_1 in LTB(u_1) which connects to v_0. Next, G-FINDER will choose a good candidate v_2 in LTB(u_2) which connects to both v_0 and v_1. Finally, G-FINDER will choose a good candidate v_3 in LTB(u_3) which connects to both v_0 and v_2.

There are several methods for choosing matching order, such as the least frequent path first [169], greedy approach to order paths [10], and the least frequent node first [127]. The general idea behind these methods is to select the node or path with the smallest number of candidate vertices first, which will make the search space compact. In this chapter, we found that these methods lead to similar performance in terms of matching accuracy for G-FINDER, and we use the greedy approach to order the paths for its computational efficiency.

3.3.4 Step 4: Subgraph-Enumeration

In this step, we search candidates in the LTBG according to the matching order from Step 3, and find the *top-k* approximate matching subgraphs which have the smallest loss function cost. During the matching process, we maintain another heap to store the k best matching results so far, and update the heap when we find a better result.

3.4 Method Details

In the following sections, we presents the details of LOOKUP-TABLE-GRAPH construction and subgraph enumeration, which are the two most complicate steps in the proposed G-FINDER (Fig. 3.2). At the end of these sections, we analyze the time and space complexity of G-FINDER.

3.4.1 LOOKUP-TABLE-GRAPH Construction

A—Lookup Table Structure. We design a novel data structure called LOOKUP-TABLE to store the information of the candidate vertices, including a candidate set, parent-child relationship, the number of intermediate vertices and the node-vertex similarity. Figure 3.4 gives an example, which illustrates the LTBs of (u_0, u_1, u_2, u_3, u_4, u_7) for the query graph in Fig. 3.3 w.r.t. the data graph in Fig. 3.1, where each LTB is visualized as a table. The black solid arrows represent the parent-child relationship between different LTBs. For example, LTB(u_1) is the parent of both LTB(u_2) and LTB(u_3). The black dash arrows represent the prior-visited neighborhood relationship. For example, there is a non-tree edge between u_3 and u_4, and LTB(u_3) is built before LTB(u_4). Therefore, there is a prior-visited neighborhood relationship between LTB(u_3) and LTB(u_4). The red solid arrows represent the parent-child relationship of candidate vertices in the data graph. For example, v_0 is the parent of both v_1 and v_3. In the LOOKUP-TABLE, the node-vertex similarity (`SIM`) and intermediate vertex number (`IVN`) are recorded. For example, there is an intermediate vertex between v_0 and v_3 in the data graph (Fig. 3.1). Therefore, the entry at the third row and the third column of LTB(u_1) is 1. When building the LTB, the candidate vertices will be sorted by their intermediate vertex number (`IVN`) and node-vertex similarity (`SIM`). During the matching process, we first choose the vertex with the high node-vertex similarity (`SIM`) and small intermediate vertex number (`IVN`) so that the corresponding linear loss function cost would be small.

B—LTB Construction. Following the root node selection (Step 1 of Fig. 3.2), we dynamically explore the data graph to select the candidates for each query node. When building LOOKUP-TABLE, we design several refinement and filtering strategies to prune unpromising candidates so as to minimize the size of candidates. These refinement strategies include node-vertex similarity, intermediate vertex number and missing node/edge tolerance. To make it easier to understand, we first introduce how to construct a single LTB, followed by how to build the entire LTBG dynamically.

Algorithm 2: LTB-Construction

Input: a data graph $\mathcal{G}$, a query graph $\mathcal{Q}$, a query node u_i and its parent u_p.
Result: the LOOKUP-TABLE of u_i over $\mathcal{G}$
Mark u_i as visited; ;
for *each visited neighboring query vertex $\hat{u}$ of u_i* **do**
  for *each vertex $\hat{v} \in C(\hat{u})$* **do**
    $S \leftarrow$ Neighbor-Expander($\hat{v}$, $L(u_i)$, $\mathcal{G}$)
Set LTB(u_i)-parent = u_p ;
for *each vertex $v_n \in C(u_p)$* **do**
  for *each vertex $n \in$ (Neighbor-Expander(v_n, $L(u_i)$, $\mathcal{G}$) $\cap$ S) and n.connection $\geq$ ($|N(u_i)|$-d)* **do**
    // d is the the number of missing edges the algorithm is willing to tolerate. ;
    Add n to LTB(u_i).adj$[v_n]$ and record IVN and SIM. ;
Sort $C(u_i)$ by IVN and SIM. ;
Return LOOKUP-TABLE of u_i ;

The key of building a single LOOKUP-TABLE is to effectively find candidate vertices. We use Fig. 3.1 as an example to explain this process. Suppose we have a partial matching $\mathcal{H}_i$, e.g., (v_0, v_1, v_4) in Fig. 3.1, where i means that there are i vertices in the partial matching (i.e., $i = 3$ in this case). The corresponding query graph is $\mathcal{Q}_i$, e.g., (u_0, u_1, u_2) in Fig. 3.1, and we want to find the matching vertex in the data graph for the next node u_j (e.g., u_3).

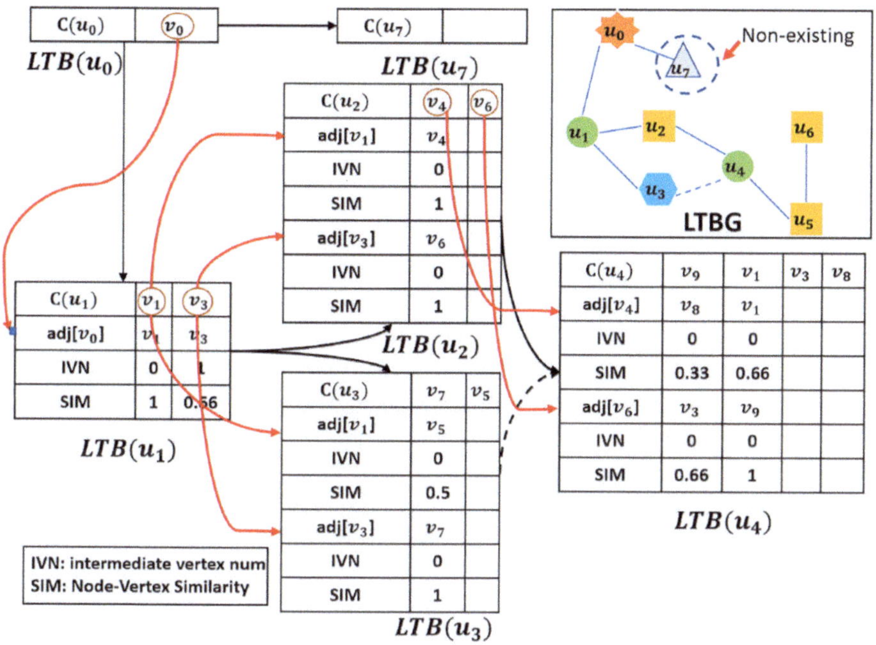

Fig. 3.4 An example of LOOKUP-TABLE-GRAPH for the data graph in Fig. 3.1 w.r.t. the query graph in Fig. 3.3. We omit the LTBs of u_5 and u_6 for brevity. An empty LTB means it is absent

Intuitively, if v_j is a good candidate of u_j, for each node $u_n \in Q_i \cap N(u_j)$, v_j must be adjacent to u_n's candidate v_n, where $v_n \in \mathcal{H}_i$. For example, v_5 is a good candidate of u_3 because it is adjacent to both v_1 and v_4. However, for approximate matching, v_j does not need to satisfy such a hard constraint if we are willing to tolerate some missing query nodes/edges. For example, in Step 4 of Fig. 3.2, (v_9, v_0, v_2, v_3) is also a good approximate matching of query graph (u_0, u_1, u_2, u_3), even though there is no direct link between v_9 and v_2 (missing edge).

Based on the ideas above, we introduce a Neighbor-Expander strategy as follows. Assuming u_p is the parent of u_j and its corresponding data vertex is v_p. Neighbor-Expander first selects all of v_p's t-hop neighbors[4] with attribute $L(u_j)$. Then, it adds these vertices into set M. For each vertex $v_m \in M$, we calculate the node-vertex similarity between v_m and u_j. If the similarity is at least s, which is a pre-defined threshold, v_m is treated as a good candidate. For example, in Fig. 3.1, v_3 is a 2-hop neighbor of v_0 and the node-vertex similarity $sim(v_3, u_1)$ is 2/3. We treat it as a good candidate if the threshold $s = 0.5$. Finally, for each v_m, we calculate how many vertices in $\mathcal{H}_i$ it connects to. If the number is bigger than $|N(u_j)| - d$, where d is the number of missing edges the algorithm is willing to tolerate, we consider it is good.

[4] $t = 2$ in this chapter for approximate subgraph matching, and users can choose a different t.

Likewise, we build a LOOKUP-TABLE by exploiting its prior-visited neighbors' LTB information. For example, assume u_i is a node in Q, and u_p is u_i's parent. When building LTB for u_i, let $N_{\text{visited}}(u_i)$ denote the set of all visited neighbors of $u_i \in Q$. Notice that all visited nodes have already built the corresponding LTBs. The candidate vertices of u_i are therefore generated from the sets of candidates in $N_{\text{visited}}(u_i)$.

The full algorithm for building a single LOOKUP-TABLE is summarized in Algorithm 2. First, for each vertex u_j in $N_{\text{visited}}(u_i)$, we iterate u_j's candidates. For each vertex $v_j \in C(u_j)$, we use Neighbor-Expander to select the good candidates v with high node-vertex similarity with respect to u_i (i.e., $sim(v, u_i) \geq s$) and insert them into vertex set S (Lines 4–6). For the example in Fig. 3.4, LTB(u_4) is built according to LTB(u_2) and LTB(u_3). The candidate sets of LTB(u_2) and LTB(u_3) are (v_4, v_6) and (v_5, v_7), respectively. Their 2-hop neighbors with attribute $L(u_4)$ are (v_1, v_8), (v_3, v_9), (v_1) and (v_3, v_9), respectively. Assume the threshold $s = 1/4$. Then, all of them are good candidates for node u_4, i.e., $S = (v_1, v_8, v_3, v_9)$. Second, we iterate each candidate vertex in $C(u_p)$, and record its parent-child relationship. For each node, we select all its t-hop neighbors in S and store these vertices in LTB(u_i).adj[v_n], which means that v_n is these vertices' common parent in the data graph. Finally, we record the number of intermediate vertices (Line 12). For the example in Fig. 3.4, the parent node of u_4 is u_2 and $C(u_2) = (v_4, v_6)$. Therefore, we have that LTB(u_4).adj[v_4]$=$ (v_8, v_1) and LTB(u_4).adj[v_6]$=$ (v_3, v_9).

Algorithm 3: LTBG-Builder

Input: a query Q, its root node u_r and the candidate vertex v_r, and a data graph $\mathcal{G}$.
Result: the LOOKUP-TABLE-GRAPH of Q over $\mathcal{G}$

```
Mark u_r as visited. ;
Initialize LTB(u_r) by adding v_r to its candidate set. ;
Push LTB(u_r) into the min heap ;
while min heap is not empty do
    currentLTB ← Pop a LTB from the min heap ;
    curQueryId ← currentLTB.Id ;
    for each un-visited neighboring query node u of curQueryId do
        nextLTB ← LTB-Construction(G, Q, u, curQueryId) ;
        LTB-heap.push(nextLTB) ;
Return LOOKUP-TABLE-GRAPH.
```

C—LTBG Construction. Besides LTB construction, another key issue of building LTBG is to select the LTBG building order for the query graph Q. The pseudo code for building LOOKUP-TABLE-GRAPH is shown in Algorithm 3. We dynamically build LTBG with a min heap to store the LTBs of all current processing nodes. First, we build LTB for the root node (Line 4), and push its LTB into the min heap (Line 5). Each time we pop a LTB from the min heap, we build LTB for its un-visited children (Lines 7, 9). Then, we push its children LTB into the min heap (Line 11). We repeat this procedure until all nodes in the query graph are processed. During this process, a subtle issue is how to pop the next LTB from the min heap. We select the LTB with the minimum $\frac{|C(u)|}{\deg(u)}$ for the following reason. If a LTB has smaller $\frac{|C(u)|}{\deg(u)}$, its children will be more likely to have less candidate vertices.

3.4.2 Subgraph-Enumeration

Algorithm 4: Subgraph-Enumeration

Input: a parameter k, the LTBG, a query graph Q, a data graph $\mathcal{G}$, the matching order M, and the linear loss function $f()$.
Result: the *top-k* approximate subgraphs

```
while H_i ← search subgraph by M do
    cost = f(Q_i, H_i) ;
    if cost > max loss function cost in the max heap then
        backtrace, stop further expanding H_i.
    if H_i can not be further expanded then
        if cost < max loss function cost in the max heap then
            Add H_i to top-k heap
Return the top-k approximate subgraphs
```

In this subsection, we describe the *top-k* search algorithm (i.e., Step 4 of Fig. 3.2). First, a new *top-k* max heap is maintained to store the best k answers that have been seen so far. During the search process, we calculate the current linear loss function cost between the partial query graph Q_i (i means that there are i nodes in the partial query graph Q_i) and the partial matching graph $\mathcal{H}_i$. Then, we compare the linear loss function cost with the maximum loss function cost in the heap. If the current loss function cost is higher than the maximum loss function cost in the max heap, we stop further expanding the current partial matching to allow earlier termination.

For example in Fig. 3.5, suppose the matching order is $< u_0, u_1, u_7, u_2, u_3, u_4, u_5, u_6 >$ and we have gotten a partial matching $\mathcal{H}_5 = (v_0, v_2, v_3, v_6, v_7)$. The corresponding partial query graph is $Q_5 = (u_0, u_1, u_2, u_3, u_7)$. If $w_1 = w_2 = w_3 = 1$, the loss function cost is 2 with one intermediate data vertex v_2 and one missing query node u_7. If this is higher than the maximum loss function cost in the max heap, we stop further expanding the current partial matching. Otherwise we continue to expand the current partial matching subgraph. The full algorithm of Subgraph-Enumeration is shown in Algorithm 4.

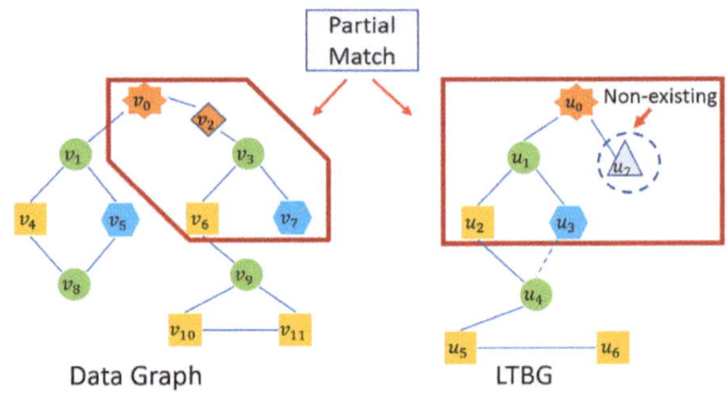

Fig. 3.5 An example of partial matching

3.4.3 Complexity Analysis

The complexity of G-FINDER mainly come from two parts: LTBG-Builder (Step 2) and Subgraph-Enumeration (Step 4). The worst-case space complexity of LTBG-Builder for a query Q over a data graph $\mathcal{G}$ is $O(|V(Q)| \times |E(\mathcal{G})|)$ and its time complexity is $O(|E(Q)| \times |E(\mathcal{G})| \times |V(\mathcal{G}|)$. For Subgraph-Enumeration, its time complexity in the worst case is $O(\prod_{i=0}^{|V(Q)|-1} |V(\mathcal{G}) - i|)$. However, the worst case only happens when the data graph is a complete graph, which is extremely unlikely for any real graph. The best case time complexity is $O(|V(Q)|)$, which happens when all nodes have distinct attributes. On average, for a data graph $\mathcal{G}$ with D attributes, its worst time complexity is $O(\prod_{i=0}^{D} \prod_{j=0}^{\frac{|V(Q)|}{D}-1} |\frac{|V(\mathcal{G})|}{D} - j|)$. Empirically, we find the total running time of G-FINDER scales near-linearly with respect to the number of nodes of the data graph and that of the query graph, for the data graph with rich attribute information.

3.5 Experiment

In this section, we conduct empirical studies to evaluate the effectiveness and efficiency of the proposed algorithm. The experiments are designed to answer the following questions:

- **Q1. Effectiveness:** How accurate is the proposed G-FINDER algorithm for subgraph matching?
- **Q2. Efficiency:** How fast and scalable is the proposed G-FINDER algorithm?

Datasets. We use 8 real-world datasets in our experiments, which are summarized in Table 3.3. `Human` and `HPRD` are two protein-protein interaction networks used by [10]. `DBLP` is a dataset which represents the relationship between authors and papers. `Flickr` shows the friendship of users on Flickr. `LastFm` contains the following relationships of users on LastFm. `AMiner` represents the academic social network, where undirected edges represent co-authorship and the node attribute vector is extracted from the number of published papers.

Table 3.3 Summary of datasets

Name	# of Nodes	# of Edges	Attribute	# of Attribute
Human	4,674	86,282	Node only	44
HPRD	9,460	37,081	Node only	307
DBLP	9,143	16,338	Node only	29
Flickr	12,974	16,149	Node only	3
LastFm	136,421	1,685,524	Node only	3
AMiner	1,274,360	4,756,194	Node only	300
PNNL-V4	22,154	460,196	Edge only	259,917
IMDB	2,932,657	11,040,263	Node & Edge	# of Nodes + # of Edges

PNNL-V4 is created by Pacific Northwest National Laboratory (PNNL). IMDB is the same movie dataset used in NeMa [64].

Baseline Methods. Five algorithms are used as the baselines, including G-Ray [144], MAGE [114], FIRST [5], FilM [34] and NeMa [64]. Source codes of the baseline methods are provided by the original authors. For the proposed G-FINDER algorithm, we have two variants based on different traversing strategies outlined in Sect. 3.2, including G-FINDER-BFS and G-FINDER-Dynamic.

Evaluation Metrics. We use the F1-Score to evaluate the matching accuracy and total running time[5] to evaluate the efficiency of different algorithms. Other alternative accuracy metrics include (1) the linear loss function defined in Sect. 3.2; (2) % Extra nodes in MAGE and FIRST; (3) % Exact Matching Nodes in FIRST. The proposed G-FINDER outperforms the baseline methods on these alternative metrics as well. Due to the space limitation, we only show the results of F1-Score.

Reproducibility. G-FINDER is implemented in C++. Experiments are conducted on a machine with an Intel Core-i7 3.20GHz CPU and 32GB memory.

3.5.1 Effectiveness Results

Here, we compare G-FINDER-Dynamic and G-FINDER-BFS against three baseline algorithms, including G-Ray, MAGE and FIRST. For the query graph, it is generated as follows. We select a connected subgraph of the data graph, and then inject a variety of different types of noise, e.g., node addition, node attribute alteration, edge deletion. More specifically, we find a induced subgraph in the data graph, and treat it as the query graph. This means that there is at least one exact matching for the query graph. Then, we will add some noises into

[5] The index construction time is smaller, compared to the query runtime.

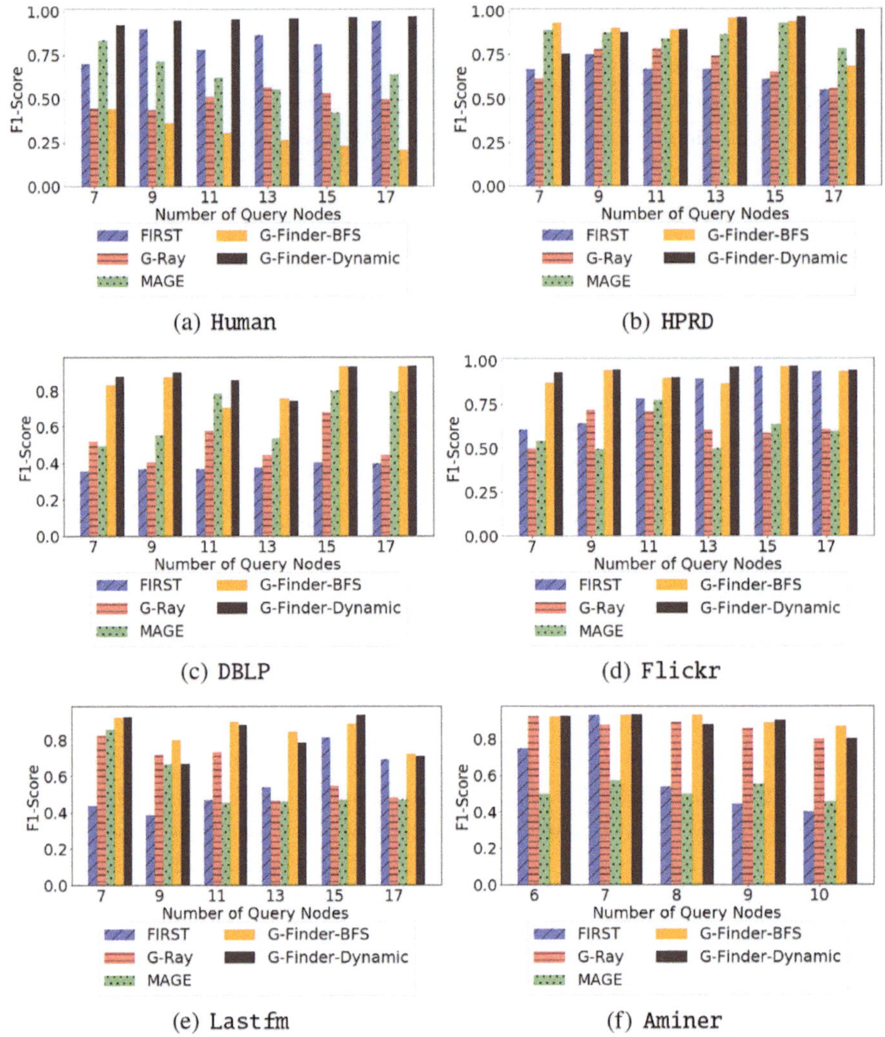

Fig. 3.6 Matching accuracy (F1-Score) comparison. Higher is better. The proposed G-FINDER-Dynamic is the rightmost bar

the query graph. For example, we randomly delete some edges in the query graph, change some nodes' attributes or add some extra nodes into the query graph. These extra nodes don't exist in the data graph. The ground truth is the induced subgraph. The main parameters are set as $k = 10$, $w_1 = w_2 = w_3 = 1$.[6] Figure 3.6 shows the F1-Score of these 5 algorithms on 6 datasets, including `Human`, `HPRD`, `DBLP`, `Flickr`, `Lastfm` and `AMiner`. The X-axis

[6] This means that G-FINDER has no bias. For example, if w_1 is larger, G-FINDER prefers to find more result nodes, but the result graph may contain a lot of intermediate nodes.

is the number of query nodes, and the Y-axis shows the F1-Scores. We can see that the proposed G-FINDER-Dynamic consistently outperforms all the three baseline methods in all cases. For example, on DBLP, G-FINDER-Dynamic is 30% better than the best baseline method (MAGE) with 7, 9 and 13 query nodes. Between the two variants of the proposed G-FINDER, G-FINDER-Dynamic wins or achieves the similar high score in almost all cases, with the only three exception on (1) HPRD with 7 query nodes, (2) Lastfm with 9 nodes and (3) AMiner with 10 query nodes. The relatively lower F1-Score of G-FINDER-BFS in Human is due to the following reason. **The noise nodes are close to the root node**. After transforming the query graph into a BFS tree, these nodes of the BFS tree do not have any candidate in the data graph. Consequently, their children nodes are ignored. This is consistent with our analysis in Sect. 3.2. That is, despite its wide usage in exact subgraph matching, BFS-tree traversing strategy might be sub-optimal for approximate subgraph matching. Rather, the proposed dynamic-tree is a much more effective traversing strategy.

Although G-FINDER is primarily designed for approximate subgraph matching, by (1) setting $t = 1$ (i.e., only considering the direct neighbors in LOOKUP-TABLE-GRAPH) and (2) only retaining the resulting matching subgraphs with a 0 linear loss function value, it can also be used for exact subgraph matching. Here, we compare G-FINDER-Dynamic with FilM [34]. It is worth pointing out that for each query node, FilM returns a candidate set which contains all the possible candidates, including potential false positives. For example, Fig. 3.7a is the result generated by FilM on PNNL-V4 dataset. The number inside each query node is the number of candidate vertices for it (e.g., a query node with 6 means that it has 6 candidates). The total number of resulting matching subgraphs by FilM are more than 100 (i.e., the multiplication of the candidate numbers for all query nodes). In this case, there

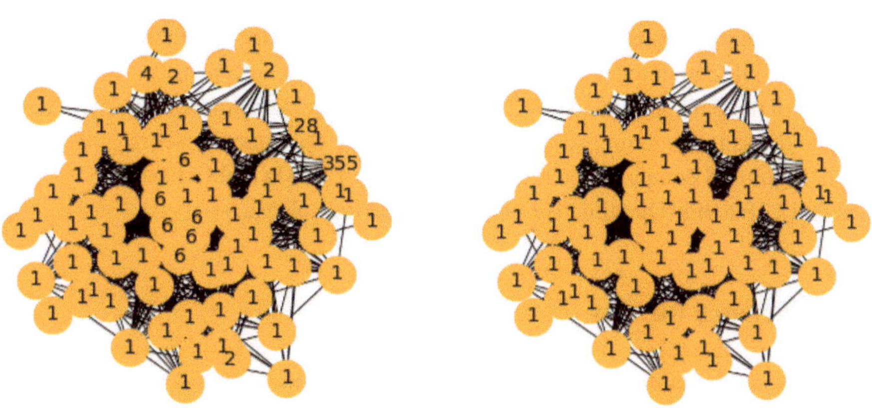

(a) FilM (b) G-FINDER

Fig. 3.7 Comparison between G-FINDER-Dynamic and FilM. Numbers are the candidate numbers for the corresponding query nodes

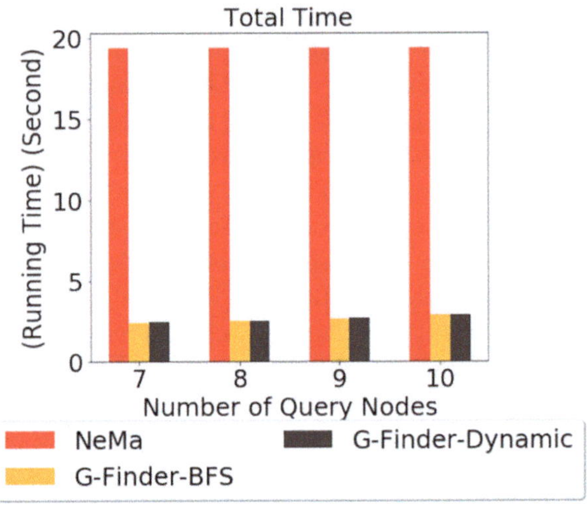

Fig. 3.8 Running time comparison: G-FINDER-Dynamic versus NeMa

is actually only one exact matching subgraph in the underlying data graph. Therefore, the F1-Score of FilM is close to 0. On the other hand, as we can see from Fig. 3.7b, the proposed G-FINDER-Dynamic generates exactly one candidate for each query node, precisely finding the only exact matching subgraph.[7]

We also compare G-FINDER-Dynamic with another algorithm NeMa [64]. NeMa is for approximate matching, which aims at finding *top-k* matching subgraphs. Different from other baseline methods, NeMa treats each node attribute value as a set of words and uses Jaccard similarity to measure the similarity between attribute values. When comparing G-FINDER-Dynamic with NeMa, to make a fair comparison, we use the same dataset (IMDB) and the same query graphs with node addition noise as NeMa. For each attribute value in IMDB, we map it to an integer value when running G-FINDER-Dynamic.

Since both the data graph and query graph have distinct attribute values, these two algorithms find the same matching subgraphs, so they have the same F1-Score (the same effectiveness). Figure 3.8 compares the total running time of NeMa and G-FINDER-Dynamic. We can see that G-FINDER-Dynamic is 6–8 times faster than NeMa.

3.5.2 Efficiency Results

Figure 3.9 shows the trade-off between the total running time and matching accuracy (measured by F1-Score) of different methods. Each point in the figure represents a pair of F1-Score and running time, and we present 5 pairs (i.e., points) for each method. As we can see, G-Ray, MAGE and G-FINDER are fastest, all of which can find matching results in a quite

[7] We also found that G-FINDER-Dynamic is faster than FilM (134 s vs. 199 s, total running time). There is only one query graph in the PNNL dataset.

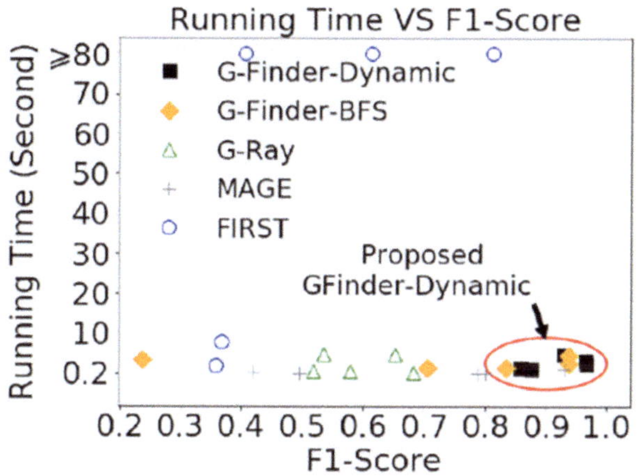

Fig. 3.9 Matching accuracy and running time trade-off

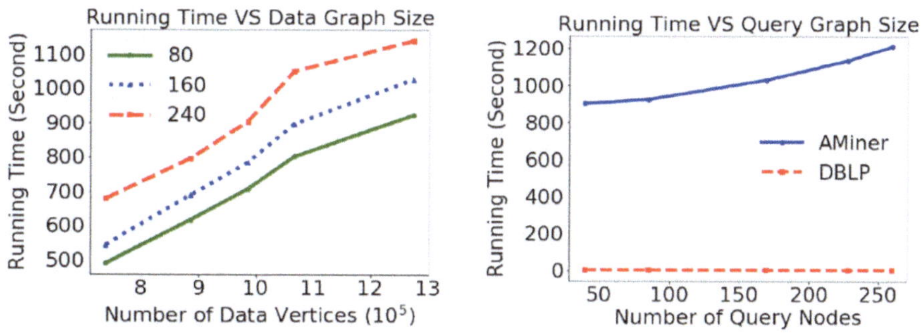

Fig. 3.10 Scalability of G-FINDER-Dynamic

short time. However, the matching accuracy (X-axis) of the proposed G-FINDER-Dynamic is much higher than other methods. In most cases, F1-Scores of G-FINDER-Dynamic are greater than 0.8.

Figure 3.10a, b show the scalability of G-FINDER-Dynamic. We can see that the running time of G-FINDER-Dynamic scales near-linearly w.r.t. the number of nodes of the data graph as well as that of the query graph.

3.6 Discussion

In this paper, we introduce a symbolic subgraph matching method that iteratively traverses the query graph to find answers in the data graph. Our method calculates the matching order based on the query structure and retrieves answers accordingly. Unlike neural network-based methods, the symbolic approach is more accurate and explainable. However, the symbolic subgraph matching method becomes time-consuming as the number of symbolic relations in the query graph increases. Moving forward, integrating symbolic reasoning with neural methods to accelerate the reasoning process is an important challenge that needs to be addressed.

Symbolic Reasoning for Inconsistency Detection Over Complete KG

4

Inconsistency detection/fact-checking for queries over a knowledge graph involve verifying the accuracy and consistency of information within the graph in response to user queries. Knowledge graphs are widely used in fact-checking because they provide crucial background knowledge that helps in verifying claims. When a claim is made, the system evaluates its authenticity by examining relevant data in the knowledge graph. In this chapter, we present our work on detecting inconsistencies in a given query using symbolic reasoning techniques applied to complete knowledge graphs. The core techniques in this chapter include symbolic subgraph extraction and similarity calculation based on random walk graph kernels.

4.1 Introduction

Knowledge graphs are ubiquitous data structure which are used to store really world entities (e.g., `Alan Turing`) and their relations (`Alan Turing`, `wasBornIn`, `United Kingdom`), which can be found in many applications [122–124]. Since the debut in 2012, several widely used knowledge graphs have been proposed, which include Yago, Wikidata, Freebase and so on. Knowledge graph reasoning which aims to discover/explain existing knowledge or infer new knowledge from existing information in the knowledge graph has emerged as an important research direction over the last few years [14, 15, 86].

Despite the great achievement in both academia and industry, most of the existing works on knowledge graph reasoning belong to the *point-wise* approaches, which perform reasoning w.r.t. *a single piece of clue* (e.g., a triple [11], a multi-hop query [86], a complex query graph [80]). For example in fact checking, given *a claim* (e.g., represented as a triple of the knowledge graph), it decides whether the claim is authentic or falsified [30, 130]. However, *comparative reasoning* [83, 84] is rarely studied. Different from point-wise reasoning (or reasoning over knowledge graph). *Comparative reasoning* over knowledge graph [83] focuses on inferring commonality and/or inconsistency with respect to multiple pieces of

L. Liu and H. Tong, *Neural Symbolic Knowledge Graph Reasoning*, Synthesis Lectures on Computer Science, https://doi.org/10.1007/978-3-032-15858-1_4

clues (e.g., multiple claims about a news article), which is a new research direction over knowledge graphs and can be widely applied to other applications, e.g., fact checking.

Comparative reasoning has many unique advantages compared with point-wise (single claim) fact checking. This is because in many real-world situations, e.g., multimodal fake news detection, single claim fact checking alone is insufficient, while comparative reasoning offers a more complete picture w.r.t. the input clues, which in turn helps the users discover the subtle patterns (e.g., inconsistency) that would be invisible by point-wise approaches. When we verify the two claims/triples at the same time, the result may be inconsistent even though each claim/triple itself is consistent if we evaluate it individually. Figure 4.1 gives an example to illustrate the power of comparative reasoning. Suppose there is a multi-modal news article and we wish to verify its truthfulness. To this end, two query graphs are extracted from the given news, respectively. One query graph contains all the information from the text, and the other contains the information from the image. If we perform point-wise reasoning to check each of these two query graphs *separately*, both seem to be true. However, if we perform reasoning w.r.t. both query graphs simultaneously, and by *comparison*, we could discover the subtle inconsistency between them (i.e., the different air plane types, the difference in maximum flying distances). In addition, comparative reasoning can also be used in knowledge graph expansion, integration and completion [83].

In this chapter, we address the problem of comparative reasoning. We mainly focus on two problems: pairwise comparative reasoning and collective comparative reasoning. To be specific, we address two key challenges as follows. We leverage graph kernel to reveal the commonality and inconsistency among input clues according to the information in the

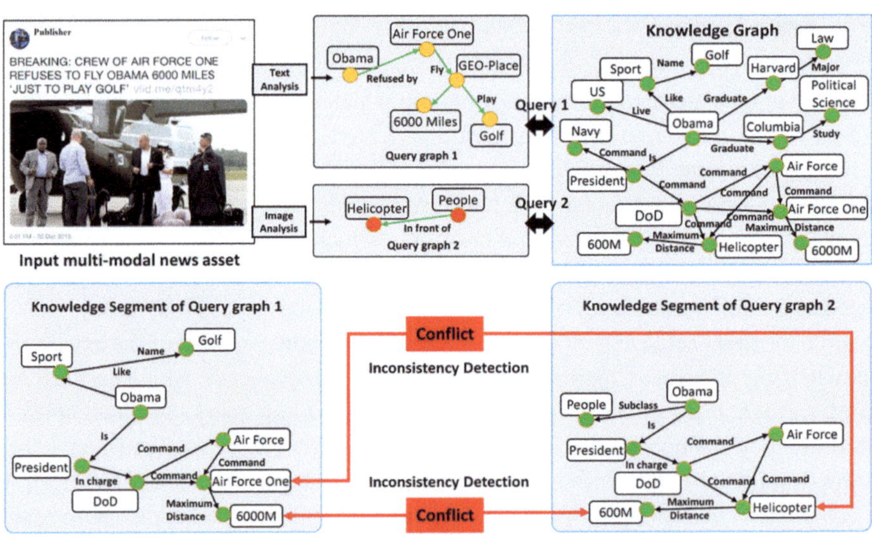

Fig. 4.1 An illustrative example of using comparative reasoning for semantic inconsistency detection. Source of the image at the top-left: [33]. The example is borrowed from [83]

background knowledge graph. We propose two different algorithms and demonstrate their effectiveness. A common building block of comparative reasoning is *knowledge segment*, which is a small connection subgraph of a given clue (e.g., a triple or part of it) to summarize its semantic context. Based on that, we present core algorithms to enable both *pairwise* reasoning and *collective* reasoning. The key idea is to use the structure and semantic information in knowledge segments to help discover vague contradictions. We perform extensive empirical evaluations to demonstrate the efficacy of our proposed methods.

4.2 Problem Definition

In this section, we first introduce the symbols that will be used throughout the chapter, then we introduce other important concepts and formally define the comparative reasoning problem.

Table 4.1 gives the main notations used throughout this chapter. A knowledge graph can be denoted as $\mathcal{G} = (V, R, E)$ where $V = \{v_1, v_2, ..., v_n\}$ is the set of nodes/entities, $R = \{r_1, r_2, ..., r_m\}$ is the set of relations and E is the set of triples. Each triple in the knowledge graph can be denoted as (h, r, t) where $h \in V$ is the head (i.e., subject) of the triple, $t \in V$ is the tail (i.e., object) of the triple and $r \in R$ is the edge (i.e., relation, predicate) of the triple which connects the head h to the tail t.

Given multiple pieces of clues, the goal of comparative reasoning is to infer commonality and/or inconsistency among them. If the given information is a pair of clues, we call it pairwise comparative reasoning or pairwise fact checking. The goal is to deduce whether these two clues are coherent or not. If the given information is a connected query graph, the goal is to detect whether there is inconsistency inside the given graph. The problem

Table 4.1 Notations and definitions

Symbols	Definition	Symbols	Definition
$\mathcal{G} = \{V^G, E^G, L^G\}$	A knowledge graph	v_i	The ith entity/node in knowledge graph
r_i	The ith relation/edge in knowledge graph	e_i	The ith given by the user
$\mathcal{Q} = \{V^Q, E^Q, L^Q\}$	An attributed query graph	KS_i	Knowledge segment i
A_i	Adjacency matrix of KS_i	$A_\times$	Kronecker product of A_1 and A_2
N^l	Diagonal matrix of the lth node attribute	$N_\times$	Combined node attribute matrix
N_i	Attribute matrix of KS_i, the jth row denotes the attribute vector of node j in KS_i	$S^{i,j}$	Single entry matrix $S^{i,j}(i, j) = 1$ and zeros elsewhere

is called collective comparative reasoning or collective fact checking. Different from traditional point-wise reasoning methods, comparative reasoning can unveil some subtle patterns which point-wise approaches may overlook. Take knowledge graph based fact checking as an example, considering two claims/triples: (`Barack Obama`, `graduatedFrom`, `Harvard University`) and (`Barack Obama`, `majorIn`, `Political Science`). Even each clue/claim is true, but if we check them together at the same time, we can see that they cannot be both true. This is because `Barack Obama` majored in `law` instead of `Political Science` when he studied at `Harvard University`. So, we might fail to detect the inconsistency between them without appropriately examining different clues/claims together.

To facilitate comparative reasoning, how to utilize the background information in knowledge graph is an important problem. If we can find a subgraph in the knowledge graph which can best express the semantic meaning of each input clue, the hidden conflicts can be easier to detect. Ideally, this subgraph should contain all the meaningful/important entities and relations in the knowledge graph which are related to the given clue. We call this subgraph *knowledge segment*, which is formally defined as follows.

Definition 4.1 (*Knowledge Segment (KS for short)*) is a connection subgraph of the knowledge graph that best describes the semantic context of a piece of given clue (i.e., a node, a triple or a query graph).

Figure 4.1 gives an example of knowledge segment. Given two clues which are two query graphs extracted from the text and image respective, their corresponding knowledge segments are shown in the bottom of the Figure. As we can see, expressing the given clues with knowledge segments can help us detect the inconsistency without difficulty.

Given the knowledge segments of multiple pieces of clues, comparative reasoning aims to infer the commonality and inconsistency among these knowledge segments to make decision. For pairwise case, the commonality refers to the common elements of these two knowledge segments. The inconsistency includes any elements that are contradicts with each other. Assuming the two given clues are two edges/triples: $E_1^Q =< \mathtt{s}_1, \mathtt{p}_1, \mathtt{o}_1 >$ and $E_2^Q =< \mathtt{s}_2, \mathtt{p}_2, \mathtt{o}_2 >$ where $\mathtt{s}_1, \mathtt{o}_1, \mathtt{s}_2, \mathtt{o}_2 \in V$ and $\mathtt{p}_1, \mathtt{p}_2 \in E$. We denote their corresponding knowledge segments as KS_1 for E_1^Q and KS_2 for E_2^Q, respectively. The commonality and inconsistency between these two knowledge segments are defined as follows.

Definition 4.2 (*Commonality*). Given two triples (E_1^Q and E_2^Q) and their knowledge segments (KS_1 and KS_2), the commonality of these two triples refers to the shared nodes and edges between E_1^Q and E_2^Q, as well as the shared nodes and edges between KS_1 and KS_2: $((V^{KS_1} \cap V^{KS_2}) \cup (V^{Q_1} \cap V^{Q_2}), (E^{KS_1} \cap E^{KS_2}) \cup (E^{Q_1} \cap E^{Q_2}))$.

Definition 4.3 (*Inconsistency*). Given two knowledge segments KS_1 and KS_2, the inconsistency between these two knowledge segments refers to any element (node, node attribute or edge) in KS_1 and KS_2 that contradicts with each other.

Different from pairwise comparative reasoning, collective comparative reasoning aims to find the commonality and/or inconsistency inside a query graph which consists of a set of inter-connected edges/triples. The corresponding definition is given below.

Definition 4.4 (*Collective Commonality*). For each edge E_i^Q in a query graph Q, let KS_i be its knowledge segment. The collective commonality between any triple pair in the query graph is the intersection of their knowledge segments.

Definition 4.5 Collective Inconsistency. For each edge E_i^Q in a query graph Q, let KS_i be its knowledge segment. The collective inconsistency refers to any elements (node or edge or node attribute) in these knowledge segments that contradict with each other.

Given the above notation and information, the problem of comparative reasoning is formal defined as:

Problem 4.1 Pairwise Comparative Reasoning:
Given: (1) A knowledge graph $\mathcal{G}$, (2) two triples E_1^Q and E_2^Q;
Output: A binary decision regarding the consistency of E_1^Q and E_2^Q.

Problem 4.2 Collective Comparative Reasoning:
Given: (1) A knowledge graph $\mathcal{G}$, (2) a query graph Q;
Output: A binary decision regarding the consistency of Q.

4.3 Knowledge Segment Extraction

In this section, we introduce how to extra knowledge segment to best express the semantic meaning of a given claim. We first introduce how to transform the knowledge graph into a relation specified weighted graph, then introduce how to extract Edge-specific Knowledge Segment and Subgraph-specific Knowledge Segment from it.

Generally speaking, given a clue (e.g., a triple or a query graph) from the user, the goal of knowledge segment extraction is to extra a subgraph which can best express the semantic meaning of the given clue. Many existing methods have been proposed to extract a concise subgraph from the source node of the querying edge to its target node in weighted or unweighted graphs. For example, multi-hop method [49], minimum cost maximum flow method [125], K-simple shortest paths based method [45] or connection subgraph [43, 68].

However, these methods do not directly apply to knowledge graphs because the edges (i.e., predicates) of a knowledge graph have specific semantic meanings (e.g., types, relations). To address this issue, we seek to convert the knowledge graph to a weighted graph by designing a predicate-predicate similarity measure for knowledge segment extraction.

4.3.1 Predicate-Predicate Similarity

In order to transform the knowledge graph into weighted graph, We propose to use a TF-IDF based method to measure the similarity between different predicates, and transfer the knowledge graph into a weighted graph whose edge weight represents the similarity between the edge predicate and query predicate. The key idea behind TF-IDF based method is that we can treat each triple in the knowledge graph and its adjacent neighboring triples as a document, and use a TF-IDF like weighting strategy to calculate the predicate similarity. For example, predicate `receiveDegreeFrom` may have neighbor predicates `major` and `graduateFrom`. These predicates should have high similarity with each other.

To be specific, we use the knowledge graph to build a co-occurrence matrix of predicates, and calculate their similarity by a TF-IDF like weighting strategy as follows. Let i, j denote two different predicates. We define the TF between two predicates as $\text{TF}(i, j) = \log(1 + C(i, j)w(j))$, where $C(i, j)$ is the co-occurrence of predicate i and j. The IDF is defined as $\text{IDF}(j) = \log \frac{|M|}{|\{i:C(i,j)>0\}|}$, where M is the number of predicates in the knowledge graph. Then, we build a TF-IDF weighted co-occurrence matrix U as $U(i, j) = \text{TF}(i, j) \times IDF(j)$. Finally, the similarity of two predicates is defined as $\text{Sim(i, j)} = \text{Cosine}(U_i, U_j)$, where U_i and U_j are the ith row and j^{th} row of U, respectively.

For the predicate-predicate similarity, suppose we want to calculate the similarity between `major` and `study`. Both `major` and `study` have only one adjacent neighboring predicate `graduate`. This means that for any predicate $i \neq \texttt{graduate}$, $U(\texttt{major}, i) = U(\texttt{study}, i) = 0$. Since $\text{E}(\texttt{graduate}) = 0$, we have $w(\texttt{graduate}) = 2\sigma(\infty) - 1 = 1$. We have $\text{TF}(\texttt{major}, \texttt{graduate}) = \text{TF}(\texttt{study}, \texttt{graduate}) = \log(1 + 1 \times 1) = 1$, and $U(\texttt{major}, \texttt{graduate}) = U(\texttt{study}, \texttt{graduate}) = IDF(\texttt{graduate}) = \log \frac{8}{4} = 1$. If we compare the two vectors, $U_{\texttt{major}}$ and $U_{\texttt{study}}$, we find that they are the same. Therefore, we have that $\text{Sim}(\texttt{major}, \texttt{study}) = 1$.

4.3.2 Edge-Specific Knowledge Segment

Edge-specific knowledge segment extraction aims at finding a knowledge segment to best characterize the semantic context of the given edge (i.e., a triple). Several connection subgraph extraction methods exist for a weighted graph, e.g., [143] uses a random walk with restart based method to find an approximate subgraph; [68] uses maximal network flow to find a subgraph and [45] aims to find a denser local graph partitions. In this chapter, after transforming the knowledge graph into a weighted graph, we find k-simple shortest paths [68] from the subject to the object of the given query edge as its knowledge segment.

4.3.3 Subgraph-Specific Knowledge Segment

Following the idea of edge-specific knowledge segment extraction, we extract a knowledge segment for each edge in the given subgraph and we call the graph which contains all the edge-specific knowledge segments subgraph-specific knowledge segment. In other words, a subgraph-specific knowledge segment consists of multiple inter-linked edge-specific knowledge segments (i.e., one edge-specific knowledge segment for each edge of the input query subgraph).

The subgraph-specific knowledge segment provides richer semantics, including both the semantics for each edge of the query graph and the semantics for the relationship between different edges of the input query graph.

4.4 Proposed Method

In this section, we introduce the technical details of comparative reasoning in KOMPARE. We first introduce the pairwise reasoning for two pieces of clues (e.g., two edges/triples), and then present the collective comparative reasoning. The main idea behind these two functions is that we use a knowledge segment to express the semantic meaning of each query triple, and use influence function to discover a set of important elements in the knowledge segments. Then, these important elements can be used to check the inconsistency.

4.4.1 Pairwise Comparative Reasoning

Pairwise comparative reasoning aims to infer the commonality and/or inconsistency with respect to a pair of clues according to their knowledge segments. Here, we assume that the two given clues are two edges/triples: $E_1^Q =< \mathtt{s}_1, \mathtt{p}_1, \mathtt{o}_1 >$ and $E_2^Q =< \mathtt{s}_2, \mathtt{p}_2, \mathtt{o}_2 >$ where $\mathtt{s}_1, \mathtt{o}_1, \mathtt{s}_2, \mathtt{o}_2 \in V^Q$ and $\mathtt{p}_1, \mathtt{p}_2 \in E^Q$. We denote their corresponding knowledge segments as KS_1 for E_1^Q and KS_2 for E_2^Q, respectively. The commonality and inconsistency between these two knowledge segments are defined as follows.

Definition 4.6 (*Commonality*). Given two triples (E_1^Q and E_2^Q) and their knowledge segments (KS_1 and KS_2), the commonality of these two triples refers to the shared nodes and edges between E_1^Q and E_2^Q, as well as the shared nodes and edges between KS_1 and KS_2: $((V^{KS_1} \cap V^{KS_2}) \cup (V^{Q_1} \cap V^{Q_2}), (E^{KS_1} \cap E^{KS_2}) \cup (E^{Q_1} \cap E^{Q_2}))$.

Definition 4.7 (*Inconsistency*). Given two knowledge segments KS_1 and KS_2, the inconsistency between these two knowledge segments refers to any element (node, node attribute or edge) in KS_1 and KS_2 that contradicts with each other.

In order to find out if these two given edges/triples are inconsistent, we first need to determine if they refer to the same/similar thing/fact. Given a pair of clues $< s_1, p_1, o_1 >$ and $< s_2, p_2, o_2 >$, we divide it into the following six cases, including

C1. $s_1 \neq s_2, s_1 \neq o_2, o_1 \neq s_2, o_1 \neq o_2$. For this case, these two clues apparently refer to different things, e.g., <Alan Turing, wasBornIn, United Kingdom> and <Google, isLocatedIn, USA>.

C2. $s_1 = s_2$ and $o_1 = o_2$. If $p_1 = p_2$, these two clues are the same. If p_1 and p_2 are different or irrelevant, e.g., $p_1 =$ wasBornIn, $p_2 =$ hasWebsite, these two clues refer to different things. However, if p_1 contradicts p_2, they are inconsistent with each other.

C3. $s_1 = s_2$ but $p_1 \neq p_2$ and $o_1 \neq o_2$, e.g., <Alan Turing, wasBornIn, Maida Vale>, <Alan Turing, livesIn, United Kingdom>.

C4. $s_1 = s_2$, $p_1 = p_2$, but $o_1 \neq o_2$, e.g., <Alan Turing, wasBornIn, Maida Vale>, <Alan Turing, wasBornIn, United Kingdom>.

C5. $o_1 = o_2$, but $s_1 \neq s_2$. For this case, no matter what p_1 and p_2 are, these two clues refer to different things.

C6. $o_1 = s_2$. For this case, no matter what p_1 and p_2 are, they refer to different things. For example, <Alan Turing, wasBornIn, United Kingdom>, <United Kingdom, dealsWith, USA>.

Among these six cases, we can see that the clue pair in C1, C5 and C6 refer to different things. Therefore, there is no need to check the inconsistency between them. For C2, we only need to check the semantic meaning of their predicates, i.e., whether p_1 contradicts p_2. For example, $p_1 =$ isFather and $p_2 =$ isSon, they are inconsistent with each other. Otherwise, there is no inconsistency between them. We mainly focus on C3 and C4 where the two clues may be inconsistent with each other even if each of them is true. For example, either <Barack Obama, graduatedFrom, Harvard University> or <Barack Obama, majorIn , Political Science> could be true. But putting them together, they cannot be both true, since Barack Obama majored in law instead of Political Science when he studied at Harvard University. In other words, they are mutually exclusive with each other and thus are inconsistent. However, queries like <Alan Turing, wasBornIn, Maida Vale> and <Alan Turing, wasBornIn, United Kingdom> are both true, because Maida Vale belongs to United Kingdom. Alternatively, we can say that United Kingdom contains Maida Vale. We summarize that if (1) the subjects of two clues are the same, and (2) their predicates are similar with each other or the same, they refer to the same thing. Furthermore, if their objects are two uncorrelated entities, it is highly likely that these two clues are inconsistent with each other.

Based on the above observations, we take the following three steps for pairwise comparative reasoning. First, given a pair of clues, we decide which of six cases it belongs to, by checking the subjects, predicates and objects of these two clues. Second, if this pair of clues belongs to C3 or C4, we need to decide whether they refer to the same thing or two different things. To this end, we first find a set of key elements (nodes or edges or node attributes) in these two knowledge segments. If most of these key elements belong to the commonality of these two knowledge segments, it is highly likely that they refer to the same thing. Otherwise, these two clues refer to different things. Third, if they refer to the same thing, we further decide whether they conflict with each other. Here, the key idea is as follows. We build two new query triples $< \texttt{o}_1, \texttt{isTypeOf}, \texttt{o}_2 >$ and $< \texttt{o}_2, \texttt{isTypeOf}, \texttt{o}_1 >$. If one of them is true, the original two triples are consistent with each other. Otherwise, they are inconsistent.

In order to find the key elements, we propose to use the influence function w.r.t. the knowledge segment similarity [173]. The basic idea is that if we perturb a key element (e.g., change the attribute of a node or remove a node/edge), it would have a significant impact on the overall similarity between these two knowledge segments. Let KS_1 and KS_2 be the two knowledge segments. We can treat the knowledge segment as an attributed graph, where different entities have different attributes. We use random walk graph kernel with node attribute to measure the similarity between these two knowledge segments [173].

$$\text{Sim}(KS_1, KS_2) = q'_{\times}(I - cN_{\times}A_{\times})^{-1}N_{\times}p_{\times} \tag{4.1}$$

where $q'_{\times}$ and $p_{\times}$ are the stopping probability distribution and the initial probability distribution of random walks on the product matrix, respectively. $N_{\times}$ is the combined node attribute matrix of the two knowledge segments $N_{\times} = \sum_{j=1}^{d} N_1^j \otimes N_2^j$ where N_i^j $(i \in \{1, 2\})$ is the diagonal matrix of the jth column of attribute matrix N_i. $A_{\times}$ is the Kronecker product of the adjacency matrices of knowledge segments A_1 and A_2. $0 < c < 1$ is a parameter.

We propose to use the influence function of $\text{Sim}(KS_1, KS_2)$ w.r.t. knowledge segment elements $\frac{\partial Sim(KS_1, KS_2)}{\partial e}$, where e represents an element of the knowledge segment KS_1 or KS_2. The element with a high absolute influence function value is treated as a key element, and it can be a node, an edge, or a node attribute. Specifically, we consider three kinds of influence functions w.r.t. the elements in KS_1, including edge influence, node influence and node attribute influence, which can be computed according to the following lemma. Note that the influence function w.r.t. elements in KS_2 can be computed in a similar way, and thus is omitted for space.

Lemma 4.1 (Knowledge Segment Similarity Influence Function [173]). *Given $Sim(KS_1, KS_2)$ in Eq. (4.1). Let $Q = (I - cN_{\times}A_{\times})^{-1}$ and $S^{j,i}$ is a single entry matrix defined in Table 4.1. We have that*

(1.) The influence of an edge $A_1(i, j)$ *in* KS_1 *can be calculated as* $I(A_1(i, j)) = \frac{\partial Sim(KS_1,KS_2)}{\partial A_1(i,j)} = cq'_{\times}QN_{\times}[(S^{i,j} + S^{j,i}) \otimes A_2]QN_{\times}p_{\times}$.
(2.) The influence of a node i *in* KS_1 *can be calculated as* $I(N_1(i)) = \frac{\partial Sim(KS_1,KS_2)}{\partial N_1(i)} = cq'_{\times}QN_{\times}[\sum_{j|A_1(i,j)=1}(S^{i,j} + S^{j,i}) \otimes A_2]QN_{\times}p_{\times}$.
(3.) The influence of a node attribute j *of node* i *in* KS_1 *can be calculated as* $I(N_1^j(i, i)) = \frac{\partial Sim(KS_1,KS_2)}{\partial N_1^j(i,i)} = q'_{\times}Q[S^{i,i} \otimes N_2^j](I + cA_{\times}QN_{\times})p_{\times}$.

Note that according to Lemma 4.1, if an element only belongs to KS_1 or KS_2, its influence function value will be 0. In order to avoid this, we introduce a fully connected background graph to KS_1 and KS_2, respectively. This background graph contains all the nodes in KS_1 and KS_2, and it is disconnected with KS_1 and KS_2. If we treat KS_1 and KS_2 as two documents, we can think of this background graph as the background word distribution in a language model.

For a given knowledge segment, we flag the top 50% of the elements (e.g., node attribute, node and edge) with the highest absolute influence function values as key elements. We would like to check whether these key elements belong to the commonality of these two knowledge segments. If most of them (e.g., 60% or more) belong to the commonality of these two knowledge segments, we say the two query clues describe the same thing. Otherwise, they refer to different things and thus we do not need to check the inconsistency between them.

If we determine that the query clues refer to the same thing, the next step is to decide whether they are inconsistent with each other. That is, given query clues $< \texttt{s}_1, \texttt{p}_1, \texttt{o}_1 >$ and $< \texttt{s}_1, \texttt{p}_2, \texttt{o}_2 >$, we need to decide whether $\texttt{o}_1$ belongs to $\texttt{o}_2$ or $\texttt{o}_2$ belongs to $\texttt{o}_1$. To this end, we build two new queries $< \texttt{o}_1, \texttt{isTypeOf}, \texttt{o}_2 >$ and $< \texttt{o}_2, \texttt{isTypeOf}, \texttt{o}_1 >$. Then, we extract the knowledge segments for these two queries, and check whether these two segments are true. If one of them is true, we say the original clues are consistent with each other, otherwise they are inconsistent. After we extract the knowledge segments for $< \texttt{o}_1, \texttt{isTypeOf}, \texttt{o}_2 >$ and $< \texttt{o}_2, \texttt{isTypeOf}, \texttt{o}_1 >$, we treat each knowledge segment as a directed graph, and calculate how much information can be transferred from the subject to the object. We define the transferred information amount as:

$$\text{infTrans}(\texttt{o}_1, \texttt{o}_2) = \max_{1 \le j \le k} \text{pathValue}(j) \tag{4.2}$$

where pathValue(j) is defined as the multiplication of the weights in the path. For an edge, its weight is the predicate-predicate similarity Sim$(\texttt{isTypeOf}, e_i)$. If a threshold T is smaller than $\max\{\text{infTrans}(\texttt{o}_1, \texttt{o}_2), \text{infTrans}(\texttt{o}_2, \texttt{o}_1)\}$, then we say $\texttt{o}_1$ belongs to $\texttt{o}_2$ or $\texttt{o}_2$ belongs to $\texttt{o}_1$. We set $T = 0.700$ in our experiment.

4.4.2 Collective Comparative Reasoning

Different from pairwise comparative reasoning, collective comparative reasoning aims to find the commonality and/or inconsistency inside a query graph which consists of a set of inter-connected edges/triples. We first give the corresponding definition below.

Definition 4.8 (*Collective Commonality*). For each edge E_i^Q in a query graph Q, let KS_i be its knowledge segment. The collective commonality between any triple pair in the query graph is the intersection of their knowledge segments.

Definition 4.9 (*Collective Inconsistency*). For each edge E_i^Q in a query graph Q, let KS_i be its knowledge segment. The collective inconsistency refers to any elements (node or edge or node attribute) in these knowledge segments that contradict with each other.

To check the inconsistency, one naive method is using the pairwise comparative reasoning method to check the inconsistency for each pair of edges in the query graph. However, this method is neither computationally efficient nor sufficient. For the former, if two clues (e.g., two claims from a news article) are weakly or not related with each other on the query graph, we might not need to check the inconsistency between them at all. For the latter, in some subtle situation, the semantic inconsistencies could *only* be identified when we collectively reason over multiple (more than two) knowledge segments. For example, given the following three claims, including (1) *Obama is refused by Air Force One*; (2) *Obama is the president of the US*; (3) *The president of US is in front of a helicopter*. Only if we reason these three claims collectively, can we identify the semantic inconsistency among them.

Based on the above observation, we propose the following method to detect the collective inconsistency.

First, we find a set of key elements inside the semantic matching subgraph. Different from pair-wise comparative reasoning, the importance/influence of an element for collective comparative reasoning is calculated by the entire semantic matching subgraph. More specifically, we first transform the query graph and its semantic matching subgraph (i.e., subgraph-specific knowledge segment) into two line graphs, which are defined as follows.

Definition 4.10 (*Line Graph* [125]). For an arbitrary graph $G = (V, E)$, the line graph $L(G) = (V', E')$ of G has the following properties: (1) the node set of $L(G)$ is the edge set of G ($V' = E$); (2) two nodes V_i', V_j' in $L(G)$ are adjacent if and only if the corresponding edges e_i, e_j of G are incident on the same node in G.

Figure 4.2 gives an example of the line graph. For the line graph $L(Q)$, the edge weight is the predicate-predicate similarity of the two nodes it connects. For the line graph $L(KS)$, the edge weight is the knowledge segment similarity by Eq. (4.1) of the two nodes it connects. The rationality of building these two line graphs is that if the semantic matching subgraph is

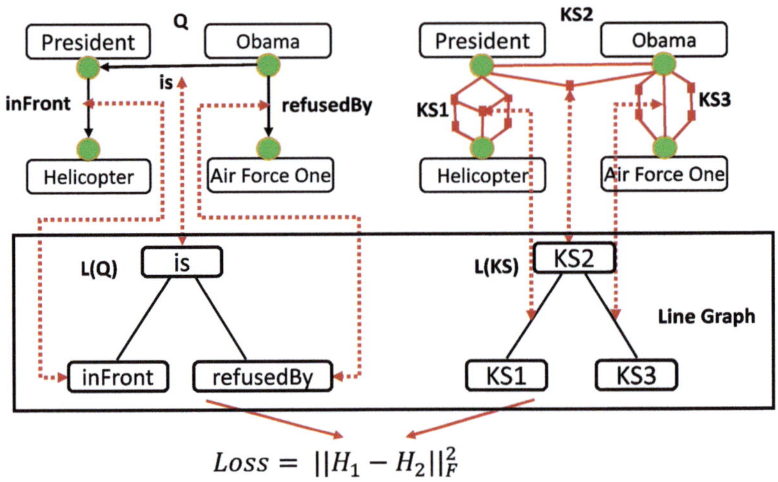

Fig. 4.2 Collective comparative reasoning workflow

a good representation of the original query graph, the edge-edge similarity in $L(Q)$ would be similar to the knowledge segment similarity in $L(KS)$.

To measure the importance of an element, we propose to use the influence function w.r.t. the distance between $L(Q)$ and $L(KS)$. We assume that a key element, if perturbed, would have a great effect on the distance $\text{Loss} = ||H_1 - H_2||_F^2$, where H_1 is the weighted adjacency matrix of $L(Q)$, and H_2 is the weighted adjacency matrix of $L(KS)$. We use the influence function $\frac{\partial \text{Loss}(H_1, H_2)}{\partial e}$, where e represents an element of the knowledge segment graph and it could be a node, an edge, or a node attribute. Lemma 4.2 provides the details on how to compute such influence functions.

Lemma 4.2 *Given the loss function* $Loss = ||H_1 - H_2||_F^2$. *Let* n, k *denote two different nodes in* $L(Q)$, *and* KS_n, KS_k *denote their corresponding knowledge segments. Let* $h_{e_{k,n}}$ *denote the weight of edge between node k and n, and* $h_{c_{k,n}}$ *denote the weight of edge between* KS_k *and* KS_n. *We have*
(1.) The influence of an edge $A_n(i, j)$ *in knowledge segment* KS_n *can be calculated as* $I(A_n(i, j)) = \sum_{k \in N(n)} -2(h_{e_{k,n}} - h_{c_{k,n}}) \frac{\partial sim(KS_n, KS_k)}{\partial A_n(i,j)}$.
(2.) The influence of a node i *in knowledge segment* KS_n *can be calculated as* $I(N_n(i)) = \sum_{k \in N(n)} -2(h_{e_{k,n}} - h_{c_{k,n}}) \frac{\partial sim(KS_n, KS_k)}{\partial N_n(i)}$.
(3.) The influence of a node attribute j *in knowledge segment* KS_n *can be calculated as* $I(N_n^j(i, i)) = \sum_{k \in N(n)} -2(h_{e_{k,n}} - h_{c_{k,n}}) \frac{\partial sim(KS_n, KS_k)}{\partial N_n^j(i,i)}$.

Second, after we find all the key elements, we check the consistency of the semantic matching subgraph according to these key elements. The steps are as follows. For each pair of knowledge segments of the semantic matching subgraph, if their key elements overlapping rate is greater than a threshold (60%), we check the consistency of this pair. Suppose the corresponding triples are $< s_1, p_1, o_1 >$ and $< s_2, p_2, o_2 >$, respectively. We check if $< s_1, \texttt{isTypeOf}, s_2 >$ or $< s_2, \texttt{isTypeOf}, s_1 >$ is true. If both of them are false, we skip this pair of clues because it does not belong to C3 or C4. Otherwise, we check if $< o_1, \texttt{isTypeOf}, o_2 >$ or $< o_2, \texttt{isTypeOf}, o_1 >$ is true. If both of them are false, we say this query graph has collective inconsistency. When checking the truthfulness of triples (e.g., $< s_1, \texttt{isTypeOf}, s_2 >$, $< s_2, \texttt{isTypeOf}, s_1 >$, $< o_1, \texttt{isTypeOf}, o_2 >$ and $< o_2, \texttt{isTypeOf}, o_1 >$), we use the same method (i.e., transferred information amount in Eq. (4.2)) as in pairwise comparative reasoning.

4.5 Experimental

In this section, we present the experimental evaluations. All the experiments are designed to answer the following two questions:

- **Q1. Effectiveness**. How effective are the proposed reasoning methods, including both point-wise methods (KOMPARE basics) and comparative reasoning methods?
- **Q2. Efficiency**. How fast are the proposed methods?

Two data graphs are used in the experiments: Yago [134][1] and Covid-19.[2] Yago [134] is a widely used knowledge graph which contains 12,430,705 triples, 4,295,825 entities and 39 predicates. The Covid-19 data graph contains three types of entities which are `Gene`, `Disease` and `Chemical`. In our experiments, we use a subset of the Covid-19 dataset which contains 55,434 core entities and 5,527,628 triples. We compare our method to two existing algorithms for fact checking (Knowledge Linker [30], short for KL, and KGMiner [130]), one link prediction algorithm (Jaccard coefficient [77]), and one algorithm for knowledge graph completion (TransE [11]). For TransE [11], we set the embedding dimension to 64 and use a margin of one and a learning rate of 0.01 for 1,200 epochs.

All the experiments are conducted on a moderate desktop with an Intel Core-i7 3.00GHz CPU and 64GB memory. The system was deployed in November 17 2020.

4.5.1 KOMPARE Basics

We evaluate the proposed predicate-predicate similarity. The top similar predicates w.r.t. `exports` by our method include `imports`, `hasOfficialLanguage`, `dealsWith`, all of which have a high similarity with `exports`. They all provide specific semantic

[1] It is publicly available at https://www.mpi-inf.mpg.de/de/departments/databases-and-information-systems/research/yago-naga/yago/downloads. We use the core version.

[2] The dataset can be found at http://blender.cs.illinois.edu/covid19/.

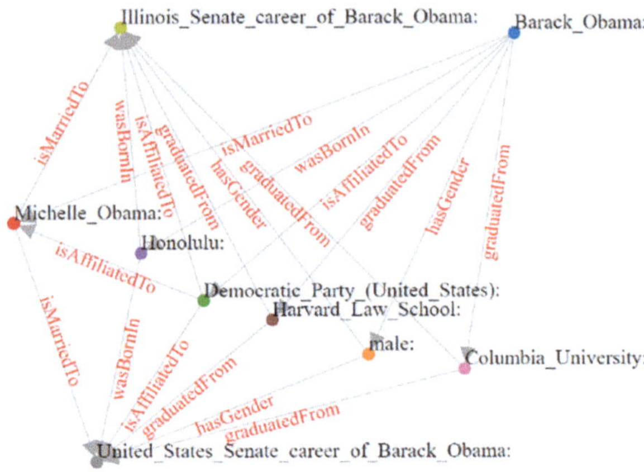

Fig. 4.3 Node-specific knowledge segment of `Barack Obama`

information about `exports`. Likewise, the top similar predicates w.r.t. `livesIn` include `wasBornIn`, `isCitizenOf`, `diedIn`, all of which are closely related to `livesIn`. These results show that the proposed TF-IDF based method can effectively measure the similarity between different predicates.

Figure 4.3 shows the node-specific knowledge segment found by our method w.r.t. the query node `Barack Obama`. We can see that the extracted knowledge segment provides critical semantics of the query node `Barack Obama`. For example, `Barack Obama` graduated from `Harvard Law School` and `Columbia University`; his wife is `Michelle Obama`; he is affiliated to the democratic party; he was the senator of united states and he was born in `Honolulu`. The result shows that node-specific knowledge segment is able to capture semantic context of the query node/entity.

4.5.2 Pair-Wise Comparative Reasoning

Here, we evaluate the effectiveness of the proposed pair-wise comparative reasoning. Ten query sets are used in the experiments. For each positive query set, it contains a set of queries which describe the true claim, while for each negative query set, it contains a set of queries which describe the false claim. For example, in query set "Birth Place", <Alan Turing, wasBornIn, Maida Vale> and <Alan Turing, wasBornIn, United Kingdom> is a positive query pair, while <Alan Turing, wasBornIn, Maida Vale> and <Alan Turing, wasBornIn, Canada> is an negative query pair. The positive queries are generated by sampling some true claims in the knowledge graph. The negative queries are generated by substituting one subject of the positive queries. The accuracy is defined as $\frac{N}{M}$ where N is the number of queries correctly classified by pair-wise comparative reasoning and M is the total number of queries. When checking the consistency of a query pair $< s_1, p_1, o_1 >$ and $< s_2, p_2, o_2 >$, because none of the base-

Table 4.2 Accuracy of pair-wise comparative reasoning

Dataset	# of queries	TransE	Jaccard	KL	KGMiner	Kompare
Family members positive	300	0.682	0.831	0.618	**0.983**	0.944
Family members negative	300	0.335	0.169	**1.000**	**1.000**	0.941
Graduated college positive	300	0.686	0.335	0.502	0.769	**0.794**
Graduated college negative	300	0.626	0.993	0.947	0.901	**0.994**
Live place positive	300	0.567	0.415	0.489	**0.834**	0.762
Live place negative	300	0.802	0.585	**0.907**	0.900	0.888
Birth place positive	300	0.590	0.435	0.537	0.698	**0.800**
Birth place negative	300	0.845	**1.000**	0.973	0.927	0.927
Work place positive	300	**0.751**	0.319	0.445	0.698	0.720
Work place negative	300	0.624	0.994	0.942	0.927	**0.995**
Mean ± *std*	–	0.651 ± 0.424	0.608 ± 0.302	0.736 ± 0.221	0.864 ± 0.105	**0.877 ± 0.095**

line methods is designed for pair-wise comparative reasoning, we use them to check each triple in the pair, if any triple is classified as false, this query pair is treated as false. Otherwise, we further check the truthness of $< \texttt{o}_1, \texttt{isTypeOf}, \texttt{o}_2 >$ and $< \texttt{o}_2, \texttt{isTypeOf}, \texttt{o}_1 >$, if one of them is classified as consistency, this query pair is treated as consistency. Table 4.2 gives the detailed results. As we can see, KOMPARE and KGMiner [130] have the highest accuracy most of the time. But KOMPARE has the highest average accuracy and the lowest variance compared with other methods.

Table 4.3 Accuracy of collective comparative reasoning

Dataset	# of queries	TransE	Jaccard	KL	KGMiner	Kompare
Birth place positive	300	0.542	0.418	0.389	0.678	**0.795**
Birth place negative	300	0.465	**0.996**	0.968	0.970	0.829
Live place positive	300	0.448	0.451	0.465	0.635	**0.989**
Live place negative	300	0.558	**1.000**	0.860	0.924	0.743
Graduated college positive	300	0.488	0.269	0.335	0.585	**0.963**
Graduated college negative	300	0.545	**0.996**	0.928	0.907	0.829
Mean ± *std*	–	0.508 ± **0.045**	0.688 ± 0.313	0.658 ± 0.265	0.783 ± 0.155	**0.858** ± 0.089

4.5.3 Collective Comparative Reasoning

We test collective comparative reasoning method on Yago dataset, using 6 query sets. Different from the queries of pair-wise comparative reasoning which only contain two edges, each query of collective comparative reasoning contains 3 edges. For example, in query set "live Place", `<Barack Obama, livesIn, Washington,D.C.>`, `<Barack Obama, is, United States Senate Barack Obama>` and `<United States Senate Barack Obama, livesIn, United States>` is a positive query triad, while `<Barack Obama, livesIn, Washington,D.C.>`, `<Barack Obama, is, United States Senate Barack Obama>` and `<United States Senate Barack Obama, livesIn, Canada>` is an negative query triad. The definition of the accuracy is the same as the previous section. Following the setting of pair-wise reasoning, when checking the consistency of the query graph $< \mathtt{s}_1, \mathtt{p}_1, \mathtt{o}_1 >$, $< \mathtt{s}_1, \mathtt{is}, \mathtt{s}_2 >$ and $< \mathtt{s}_2, \mathtt{p}_2, \mathtt{o}_2 >$, we use baseline methods to check the truthness of this query triad, if any edge is classified as false, this query triad is treated as false. Otherwise, we further check the truthness of $< \mathtt{o}_1, \mathtt{isTypeOf}, \mathtt{o}_2 >$ and $< \mathtt{o}_2, \mathtt{isTypeOf}, \mathtt{o}_1 >$, if one of them is classified as true, this query pair is treated as consistency. Table 4.3 gives the detailed results. As we can see, Jaccard [77] prefers to classify all queries as inconsistency and has the largest variance. TransE [11] has the lowest variance, but its average accuracy is very low. KOMPARE has the highest accuracy most of the time. It also has the highest average accuracy, and the second lowest variance.

Table 4.4 Accuracy of collective comparative reasoning for Covid-19

Dataset	# of queries	TransE	Jaccard	KL	KGMiner	Kompare
Positive	36	0.667	0.611	**1.000**	0.694	**1.000**
Negative	36	0.528	0.361	0.722	0.553	**0.863**
Average accuracy	–	0.598 ± 0.071	0.486 ± 0.126	0.861 ± 0.138	0.623 ± 0.071	**0.932 ± 0.063**

We further provide experimental results on Covid-19 dataset. We use queries which contain connections between drugs and genes/chemicals related to Covid-19.[3] Among all these queries, we use queries which contain less than 8 nodes, and treat them as positive queries. For each of the positive queries, we randomly select one node inside the query and substitute it with a randomly selected entity in the data graph, and treat the new query as the negative query. For all the baseline methods, we use them to check all the edges inside the query, if any edge is classified as false, the whole query is treated as false. Table 4.4 shows the accuracy of different methods. As we can see, KOMPARE has the highest accuracy on both the positive and negative datasets, it also has the highest average accuracy and the lowest variance compared with other baseline methods.

4.5.4 KOMPARE Efficiency

The runtime of knowledge segment extraction depends on the size of the underlying knowledge graphs. Among the three types of knowledge segments, subgraph-specific knowledge segment is most time-consuming. Figure 4.4a shows that its runtime scales sub-linearly w.r.t. the number of nodes in the knowledge graph. Different lines show the runtime w.r.t. different query graph size. Figure 4.4b shows the runtime of comparative reasoning, where 'Pair-wise' refers to the pairwise comparative reasoning, and the remaining bars are for collective comparative reasoning with 3, 4 and 5 edges in the query graphs respectively. Note that the runtime of comparative reasoning only depends on the size of the the corresponding knowledge segments which typically have a few or a few tens of nodes.

4.6 Discussion

Symbolic knowledge graph fact-checking verifies claims using patterns in the knowledge graph, ensuring clarity and transparency. However, it depends heavily on the graph's completeness and accuracy, which can reduce reliability if the data is incomplete or incorrect. Combining symbolic methods with neural networks can mitigate these challenges by lever-

[3] The query graphs can be found at http://blender.cs.illinois.edu/covid19/visualization.html.

Fig. 4.4 Runtime of KOMPARE

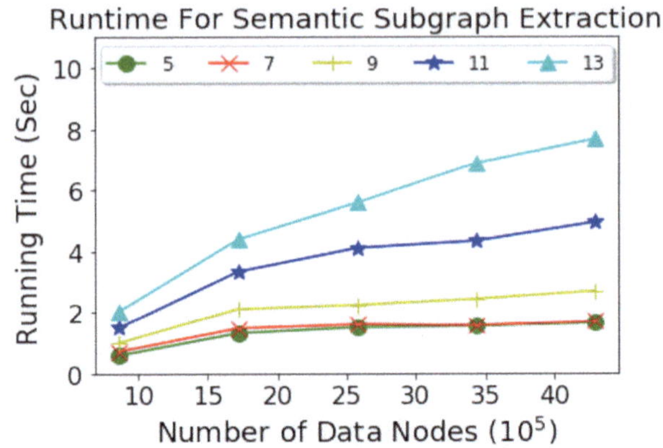

(a) Subgraph-specific KS extraction

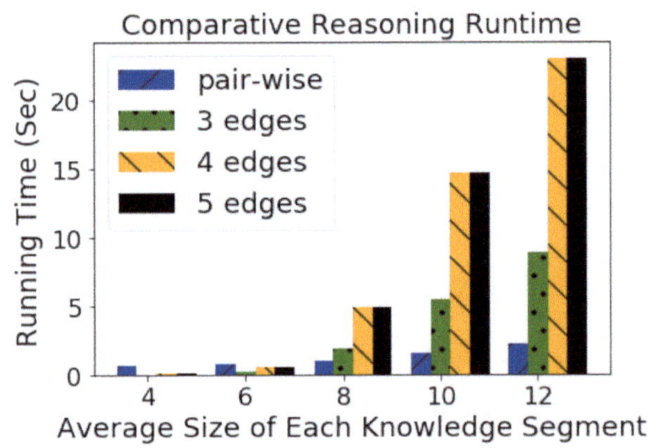

(b) Comparative reasoning

aging their respective strengths. It can enhance efficiency while preserving the interpretative power of symbolic reasoning. It represents a crucial direction for advancing more robust fact-checking systems in the future.

Accurate Query Answering with Neural Symbolic Reasoning Over Incomplete KG

5

Reasoning over incomplete knowledge graphs refers to the process of inferring logical conclusions and making accurate predictions based on limited and potentially fragmented information within a knowledge graph. In such scenarios, the graph may lack crucial triplets that are essential for comprehensive understanding and analysis. Despite these limitations, reasoning techniques aim to fill in the gaps and derive meaningful insights by leveraging the available information, patterns, and relationships within the graph. Reasoning over incomplete knowledge graphs is crucial in various domains, including information retrieval, question answering, and fact checking systems, where dealing with incomplete information is inevitable.

5.1 Introduction

A knowledge graph is a graph structure that contains a collection of facts, where nodes represent real-world entities, events and objects, and edges denote the relationship between two nodes. Since its debut in 2012, a variety of knowledge graphs have been generated, including Freebase, Yago, Wikidata and so on. The applications of knowledge graphs are numerous, ranging from network alignment [161], computational fact-checking [83, 88, 93] to recommendation [39]. The applications of knowledge graphs are numerous, ranging from network alignment [161], computational fact-checking [83, 88, 93] to recommendation [39]. Among these applications, multi-hop question answering over knowledge graph (Muti-hop KGQA for short) aims to answer natural language questions with the help of knowledge graphs. Knowledge graph completion (KGC), on the other hand, seeks to infer missing facts based on existing information in KG.

Recently, KGQA and KGC tasks have attracted great attention from both the academia and the industry, and a multitude of algorithms have been proposed. For example, for KGC, TransE [11] models the relation as a linear transition of points in the embedding space, Com-

L. Liu and H. Tong, *Neural Symbolic Knowledge Graph Reasoning*, Synthesis Lectures on Computer Science, https://doi.org/10.1007/978-3-032-15858-1_5

plEx [147] embeds relations and entities in complex space and makes predictions according to an energy function. For multi-hop question answering, Pullnet [135] first extracts question specific subgraphs, and then performs multi-hop reasoning on the extracted subgraph via graph neural networks to find answers, EmbedKGQA [126] uses a pre-trained BERT model to map natural language questions to relation embeddings and finds answers by ComplEx [147]. We note that some work [126] treats the knowledge graph completion task as a single-hop knowledge graph question answering task due to their interchangeable properties. For example, the task of predicting the answer for triple (`Interstellar`, `hasGenre`, ?) in the KGC task could be transformed to answer natural language question "*What is the genre of Interstellar?*".

Despite the potential close relationship between multi-hop KGQA and KGC tasks [126], existing works usually treat them as two separate tasks without considering their reciprocal benefits. That is, most existing multi-hop KGQA methods have implicitly assumed the background knowledge graph is complete [135, 136], whereas the existing KGC methods only exploit the existing information of the input incomplete KGs, e.g., RESCAL [107] and ComplEx [147].

Different from existing methods, we envision that knowledge graph completion (KGC) and knowledge graph question answering (KGQA) are inherently complementary with each other due to the following reasons. First, *(R1) KGQA helps KGC.* This is because new knowledge could be inferred from the KGQA task, which in turn could be used to complete the knowledge graph. For example, given the knowledge graph in Fig. 5.1, if we want to answer "*which year was A Song of Love released?*", there is no way this question could be answered because `A song of Love` only has two relations around it: `direct` and `hasGenre`. Even a human could not answer this question only based on the existing knowledge graph. However, if we are provided the answer (`1950`) to the question "*which years were all the films directed by Jean Genet released?*", we can infer that the release year of `A song of Love` is 1950 because `Jean Genet` only directed one film in his life. This suggests that KGQA could indeed help KGC. Second, *(R2) KGC helps KGQA.* This is because KGC could help improve the performance of KGQA by providing a KG with more complete knowledge

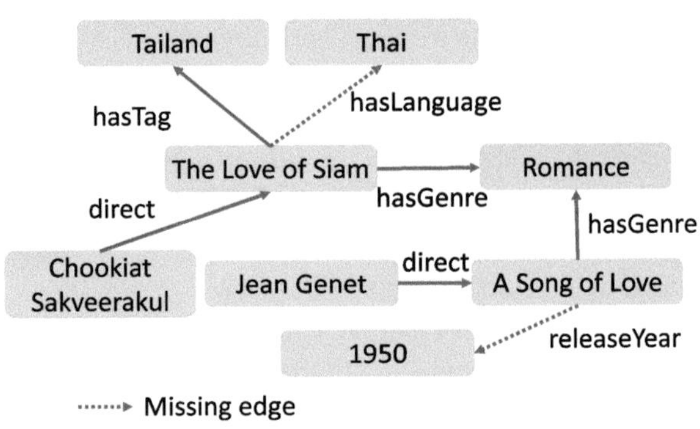

Fig. 5.1 An illustrative example of incomplete knowledge graph. The dashed line represents missing links. The solid lines are relations in the existing knowledge graph

triples of high quality. For example, because the movie `The Love of Siam` is linked to `Tailand` via the `hasTag` relation, an ideal KGC model can infer that the movie might be intended for the `Thai` audiences and augment the knowledge graph with appropriate language information for the movie. Subsequently, the question "*what is the language of the film The Love of Siam*" can then be trivially answered.

Armed with this key insight, we propose to jointly address multi-hop KGQA and KGC tasks. We formulate it as a multi-task learning problem. First, in order to leverage multi-hop KGQA for the KGC task, we propose an encoder-decoder-based model which transforms the natural language questions into relation paths to facilitate KGC. Second, in order to leverage KGC for multi-hop KGQA, we let multi-hop KGQA and KGC share both the embedding space and the answer scoring module, which allows them to automatically share latent features and reinforce each other. The experimental results on several real-world datasets demonstrate that the proposed BiNet consistently achieves state-of-the-art performance in both tasks.

5.2 Problem Definition

Table 5.1 gives the main notations used throughout this chapter. A knowledge graph can be denoted as $\mathcal{G} = (\mathcal{V}, \mathcal{R}, \mathcal{L})$ where $\mathcal{V} = \{v_1, v_2, \ldots, v_n\}$ is the set of nodes/entities, $\mathcal{R} = \{r_1, r_2, \ldots, r_m\}$ is the set of relations and $\mathcal{L}$ is the list of triples. Each triple in the knowledge graph can be denoted as (h, r, t) where $h \in \mathcal{V}$ is the head (i.e., subject) of the triple, $t \in \mathcal{V}$ is the tail (i.e., object) of the triple and $r \in \mathcal{R}$ is the edge (i.e., relation, predicate) of the triple which connects the head h to the tail t. The embedding of a node or relation is represented by bold lowercase letters, e.g., $\mathbf{e}_i$, $\mathbf{r}_i$.

Table 5.1 Notations and definitions

Symbols	Definition
v_i	The ith entity/node in knowledge graph
r_i	The ith relation/edge in knowledge graph
$\mathbf{e}_{v_i}$ / $\mathbf{e}_i$	The embedding of node v_i
$\mathbf{r}_i$	The embedding of relation r_i
$\overline{Q}$	The multi-hop natural language question training set
v_Q	The topic entity in question
$\mathbf{e}_Q$	The embedding of topic entity v_Q
$\mathbf{Q}$	The embedding of natural language question Q
w_i	The ith word in Q
P	The path decoded from question embedding
$P[i]$	The ith relation in path P

Given a knowledge graph $\mathcal{G} = (\mathcal{V}, \mathcal{R}, \mathcal{L})$ and a natural language question Q which contains a topic entity $v_Q \in \mathcal{V}$ and a sequence of words $Q = (w_1, w_2, \ldots, w_{|Q|})$, multi-hop question answering over knowledge graph aims to identify a set of nodes $A_Q \subseteq \mathcal{V}$ to answer this question. Following the definition in [126], we assume that all the answer entities exist in the knowledge graph and each question in multi-hop KBQA only contains a single topic entity $v_Q \in \mathcal{V}$ and v_Q is given. For example, `The Love of Siam` is the topic entity of "*what is the language of the film The Love of Siam*". Ideally, each question can be mapped to a unique path $P = (r_1, r_2, \ldots, r_{|P|})$ in the knowledge graph.

Knowledge graph completion intends to infer missing facts/triples based on existing information in the knowledge graph. Typically, KGC contains three kinds of sub-tasks: (1) given a partial triple $(h, r, ?)$, predict the corresponding tail entities; (2) given a partial triple $(?, r, t)$, predict the corresponding head entities; (3) given a head entity and a tail entity, predict the relationship between them. In this chapter, we only consider the first sub-task. When predicting the tail entities, the KGC method will produce a probability score for each entity in the knowledge graph, the probability score of a candidate entity v denotes how likely the triple (h, r, v) is true.

Many real-world knowledge graphs are incomplete. That is, some key information may not exist in the input knowledge graph. Performing KGQA on an incomplete knowledge graph could lead to wrong answers. However, if the knowledge graph is complete, the KGQA task is more likely to find the correct answers. On the other hand, completing an existing knowledge graph without any extra information might be hard. However, the question and answer pairs in the KGQA task could provide auxiliary information to help complete the knowledge graph. Based on this observation, we aim to jointly handle KGC and multi-hop KGQA, which can be formally defined as follows.

Proposition 5.1 *Jointly multi-hop KGQA and KGC*

Given: *(1) A set of training triples of the KGC task, (2) a set of training multi-hop natural language questions of the KGQA task;*

Output: *(1) The answer for test triples of the KGC task, and (2) the* top-k *answers for test multi-hop natural language questions of the KGQA task.*

5.3 Proposed Method

5.3.1 Model Overview

Multi-hop question answering over knowledge graph (Multi-hop KGQA) can be cast as an entity seeking problem on knowledge graph $\mathcal{G}$ by translating Q into a query path and traveling the knowledge graph to find answers. However, in many real-world cases, the knowledge graph is often incomplete. Thus, attempting to find the answer set A_Q directly on $\mathcal{G}$ by path traverse or subgraph matching [81] is impractical. Knowledge graph completion (KGC), on the other hand, often suffers from *knowledge famine* (i.e., insufficient information) or key missing information, which makes it impossible to complete the knowledge graph. For example, in Fig. 5.1, if the relation `releaseYear` between `A Song of Love` and `1950` is missing, this missing information is impossible to be completed without extra information. As mentioned in Sect. 5.2, the multi-hop KGQA and KGC are inherently complementary with each other. On one hand, the information in the natural language question and its corresponding answers can help the knowledge graph completion task. On the other hand, a more complete knowledge graph could potentially improve the KGQA accuracy. To mutually reinforce these two tasks, our model is designed to bear the following two properties. First, to help the KGC task, we use an encoder-decoder model to transform the natural language query to a path that contains one or multiple relations. Second, in our model, both KGQA and KGC leverage the same embedding space and the answer scoring module to automatically share latent features.

Figure 5.2 shows the framework of our proposed model. Given a multi-hop natural language question with a topic entity, the proposed BiNet first generates a question's embedding **Q** via a pre-trained BERT [37] model, and then uses a decoder to generate a relation path which is used by both KGQA and KGC tasks. In order to generate a high-quality path, BiNet uses a probability model to choose the path with the highest probability in the knowledge graph that best interprets the question context. After obtaining the path, the answer scoring module ranks all the nodes and selects k candidates which are most likely to answer the question. Finally, the topic entity v_Q, the path P, and k candidate nodes are treated as the input of a Transformer [149] to generate the final output of the KGQA task. In addition, the decoded path P is used by the answer scoring module to help the knowledge graph comple-

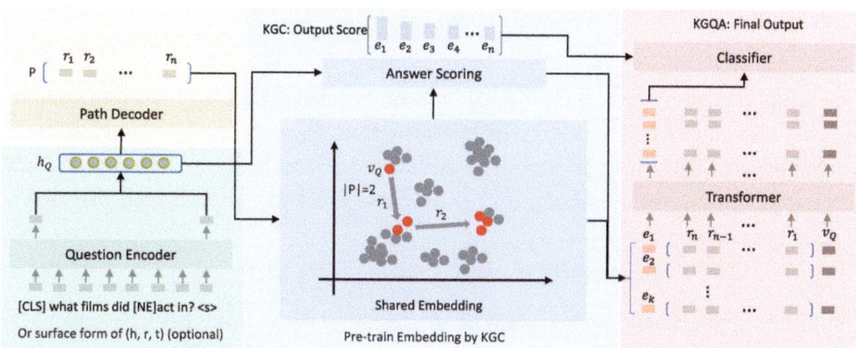

Fig. 5.2 Architecture of our proposed BiNet Model. The gray part corresponds to Sect. 5.3.3 which is used by both the KGQA and KGC tasks. The red parts are designed for the KGQA task, while the yellow part is intended for the KGC task. The green part will be used by KGQA and KGC tasks. The shared embedding space contains pre-trained embeddings of entities and relations by the training data of KGC. Best viewed in color

tion task. The overall model can be optimized in an end-to-end manner by combining the loss of different parts.

5.3.2 Question Encoder-Decoder

A—Preprocessing Text for the Question Encoder. The proposed BiNet decodes a sequence of relations between the topic entity v_Q and an answer set A_Q in a natural language sentence $Q = (w_1, w_2, \ldots, w_{|Q|})$ where $v_Q \in \mathcal{V}$ and $A_Q \subseteq \mathcal{V}$. Intuitively, each question context Q could be mapped to a relation path in the knowledge graph distinctively. Therefore, we train an encoder to represent the context as a vector.

To mitigate the noise brought by the surface forms of the entities, we introduce a special token [NE] to mask the topic entity inside the question context/surface form (e.g., "*Who starred Interstellar?*" becomes "*Who starred [NE]?*"). Masking the topic entity prevents the encoder from memorizing the surface forms of the entity and helps it generalize to similar questions involving other entities. Besides this, we add two indicator tokens ([CLS] and $< s >$) to the beginning and end of the question context to signify its boundary.

B—Question Encoder. Given the processed question context, we first pass it through a pre-trained BERT [37] to extract contextual embeddings for each token[1]:

$$\left[\mathbf{h}_{CLS}, \mathbf{w}_1, \ldots, \mathbf{w}_{|Q|}, \mathbf{h}_\mathbf{s}\right] = \text{BERT}([CLS], w_1, \ldots, w_{|Q|}, < s >) \tag{5.1}$$

where $\mathbf{h}_{CLS}$ is the embedding of the [CLS] token and $\mathbf{h}_\mathbf{s}$ is the embedding of the $< s >$ token. The final question embedding is obtained from the combination of $\mathbf{h}_{CLS}$ and $\mathbf{h}_\mathbf{s}$ as below, where FFN is a feed forward neural network, and | indicates concatenation.

$$\mathbf{h}_Q = \text{FFN}([\mathbf{h}_{CLS}|\mathbf{h}_\mathbf{s}]) \tag{5.2}$$

C—Question Decoder. Given the question context embedding $\mathbf{h}_Q$, the proposed BiNet decodes $\mathbf{h}_Q$ and generates a sequence of relations $P = (r_1, r_2, \ldots, r_n)$ using a Long Short-Term Memory (LSTM) [54] model. The initial hidden state $\mathbf{h}_0$ and initial cell state $\mathbf{c}_0$ are obtained from question embedding $\mathbf{h}_\mathbf{Q}$ by passing it through two feed forward neural networks, separately.

$$\mathbf{h}_0 = \text{FFN}_h(\mathbf{h}_Q) \qquad \mathbf{c}_0 = \text{FFN}_c(\mathbf{h}_Q) \tag{5.3}$$

The initial input embedding $\mathbf{x}_0$ could be the question embedding $\mathbf{h}_\mathbf{Q}$ or a zero vector. At time step t, the hidden state $\mathbf{h}_t$ will be passed through a feed forward neural network with batch normalization and dropout followed by a Softmax function to obtain the score of each relation to form a score vector:

[1] For dataset MetaQA, the surface forms of KGC training triples are treated as the input questions.

$$\mathbf{h}_t = \text{LSTM}(\mathbf{h}_{t-1}, \mathbf{c}_{t-1}, \mathbf{o}_{t-1}) \tag{5.4}$$

$$\mathbf{a}_t = \text{softmax}(\text{MLP}(\mathbf{h}_t)) \tag{5.5}$$

The new input embedding $\mathbf{x}_{t+1}$ of the decoder at step $t + 1$ is the weighted sum of all relation embedding of step t.

$$\mathbf{x}_{t+1} = \sum_i \mathbf{a}_t(i)\mathbf{r}_i \tag{5.6}$$

where $\mathbf{a}_t(i)$ is the i-th element of $\mathbf{a}_t$. The final prediction of each step is the relation with the highest score.

D—Training Question Encoder-Decoder. Given a natural language question with its answer set, we want to map the question context to its correct relation path. The topic entity v_Q and answer entities can be identified in the knowledge graph according to their surface forms. However, there might exist many paths between them. To identify the correct path, we find all the k-shortest paths between each entity pair (v_Q, v_i) where $v_i \in A_Q$. We treat all these shortest paths as potentially correct path candidates. We use Bayes' Rule to infer the probability of whether the shortest path is the correct mapping of the question context. First, all the questions in the training set can be divided into m groups: $S_1, S_2, \ldots, S_m$ where m is the number of unique question contexts after masking the topic entity. Each group has a corresponding answer pool $PL_i = \{A_{Q_j} | Q_j \in S_i\}$. Each answer entity $v_i \in A_{Q_j}$ has a corresponding candidate path set $PC(Q_j, v_i) = \{P_i | (v_{Q_j}, P_i, v_i) \in \mathcal{G}\}$. If we assume the occurrence of different paths in PC as i.i.d. random variables, the probability that a path interprets the question context can be expressed as

$$Pr(P_i | S_j, \theta) = \frac{\sum_{Q_j \in S_j} Pr(P_i, Q_j | \theta)}{|PL_j|} \tag{5.7}$$

where $|PL_j|$ is the number of answer sets in PL_j, $Q_j \in S_j$ denotes a specific question Q_j (e.g., "*Who starred Interstellar?*") belongs to S_j (e.g., "*Who starred [NE]?*") and $Pr(P_i, Q_j | \theta)$ is the probability that $P_i \in PC(Q_j, v_i)$ for any $v_i \in A_{Q_j}$, which is defined as follows, where $\mathbb{K}()$ is the indicator function.

$$Pr(P_i, Q_j | \theta) = \frac{\sum_{v_i \in A_{Q_j}} |PC(Q_j, v_i)|^{\mathbb{K}(P_i \in PC(Q_j, v_i))}}{|A_{Q_j}|} \tag{5.8}$$

After calculating the probability for all the potential paths, we choose the path with the highest probability as the answer. If multiple paths have the highest probability, we treat all of them as correct answers. When training the question encoder and decoder, we use binary cross-entropy loss which is defined as:

$$\mathcal{L}(\hat{P}, P) = \frac{1}{N}\sum_{i=1}^{N}\sum_{j=1}^{|P|} \mathbb{1}(P[j] = r_i)\log(Pr(r_i|\mathcal{M})) \tag{5.9}$$

$$+ (1 - \mathbb{1}(P[j] = r_i))\log(1 - Pr(r_i|\mathcal{M})) \tag{5.10}$$

where $\mathcal{M}$ is the parameters of the question encoder-decoder model, N is the number of relations in $\mathcal{G}$. P is the path obtained by Eq. (5.8). $\hat{P}$ is the output of path decoder. Note that the output of path decoder at time step t is a probability distribution over all relations in $\mathcal{G}$.

5.3.3 Answer Scoring

When answering multi-hop natural language questions, traditional knowledge graph traversal or subgraph matching method [120] is infeasible due to the incompleteness of the knowledge graph. When using learning-based methods to solve this problem, we aim to find a function $f()$ which takes in the topic entity v_Q, the path $P = (r_1, r_2, \ldots, r_{|P|})$, the knowledge graph $\mathcal{G}$ and a candidate entity v to output a score which is used to denote how likely it is to travel from v_Q, according to path P, to reach v.

A—Background. If an algorithm satisfies the transitivity property of knowledge graph, the operation on path P can be expressed as

$$\mathbf{p} = \mathbf{r}_1 \star \mathbf{r}_2 \star \cdots \star \mathbf{r}_n \tag{5.11}$$

where $\star$ is a composition operation and $\mathbf{p}$ is the embedding of P. According to different designs, the composition operation can have different forms. For example, in TransE [11], each relation represents a linear translation operation in the embedding space, so the path representation can be obtained by:

$$\mathbf{p} = \sum_{r_i \in P} \mathbf{r}_i \tag{5.12}$$

In RotaE [138], each relation represents a rotation in the complex space, and the composition operation is the Hadamard (i.e., element-wise) product which means

$$\mathbf{p} = \mathbf{r}_1 \odot \mathbf{r}_2 \odot \cdots \odot \mathbf{r}_n \tag{5.13}$$

However, naively using Eq. (5.11) to calculate the path embedding and finding answers may suffer from low accuracy. This is because noise often exists in the embedding space, and with the increase of the path length, the cascading error will become larger [60, 85]. So, it is necessary to use the intermediate candidates to adjust the search process.

B—Probabilistic Reasoning Model. We address this issue by using a probability model Θ. Considering a relation sequence $P = (r_1, \ldots, r_{|P|})$ originated from topic entity v_Q, the model predicts the likelihood $\Theta : (r, v) \rightarrow [0, 1]$ of following a certain edge in a relation sequence from v_Q to any node in A_Q. Specifically, we compute the likelihood of v by multiplying the likelihood of all intermediate steps traversed by P in $\mathcal{G} = (\mathcal{V}, \mathcal{R}, \mathcal{L})$ by:

$$Pr(v|P, v_Q, \mathcal{G}) \propto \prod_{i=1}^{|P|} \Theta(r_i, v_i | P_{1\rightarrow i-1}, v_Q, \mathcal{G}) \tag{5.14}$$

where $P_{1\rightarrow r_i}$ is the subsequence of P up to the i-th relation and $P_{1\rightarrow 0}$ means empty set. Finding the best answer is equivalent to maximizing the probability function. A naive way to find answers is by choosing the intermediate entity v_i which can maximize the probability at each time step. However, due to the incompleteness of the knowledge graph, it may fail to find the correct answer. Iterating all the intermediate candidates can make sure to find the correct answer with a high probability, but it could hamper the efficiency. In order to strike a good balance between effectiveness and efficiency, as well as mitigate the cascading error [60], at each step, we select the *top-k* candidates with maximum likelihood $Pr(v|P_{1\rightarrow i-1}, v_Q, \mathcal{G})$. This can be done by an efficient search algorithm, such as beam search starting from v_Q. In the last step, we choose the candidate with the highest probability.

$$o = \max_{v_i \in \mathcal{V}}(Pr(v_i|P, v_Q, \mathcal{G})) \tag{5.15}$$

Note that, the goal of probability model Θ is to estimate how likely an entity is the answer. It is used by both the KGQA task and the KGC task to allow them automatically share information. Many methods could be used to model Θ, e.g., bilinear transformation [162], convolutional networks [36] and so on. In our experiment, we use ComplEx [147] for its simplicity. Given $v_h, v_t \in \mathcal{V}$ and $r_i \in \mathcal{R}$, and their embedding $\mathbf{e}_h, \mathbf{e}_i, \mathbf{r}_i$, the probability score of v_t that is reachable from v_h by r_i is calculated by:

$$Pr(v_t|r_i, v_h, \mathcal{G}) = Re(< \mathbf{r}_i, \mathbf{e}_h, \bar{\mathbf{e}}_i >) \tag{5.16}$$

where $Re()$ is the real part of ComplEx [147] output.

C—Connection with Existing Methods. Most existing algorithms which satisfy the transitivity property of knowledge graph are special cases of our model. For example, in TransE [11], each relation represents a linear translation operation in the embedding space, so the probability score of (r_i, v_t) can be calculated by:

$$\Theta(r_i, v_t|P_{1\rightarrow i-1}, v_Q, \mathcal{G}) \propto ||\mathbf{e}_{v_Q} + \sum_{r_j \in P_{1\rightarrow i}} \mathbf{r}_j - \mathbf{e}_t||_2^{-\mathbb{1}(i=|P|)} \tag{5.17}$$

and similarly in RotatE [138], the probability score of (r_i, v_t) can be calculated by:

$$\theta(r_i, v_t|P_{1\rightarrow i-1}, v_Q, \mathcal{G}) \propto ||\mathbf{e}_{v_Q} \odot \mathbf{r}_1 \odot \ldots \odot \mathbf{r}_i - \mathbf{e}_t||_2^{-\mathbb{1}(i=|P|)} \tag{5.18}$$

5.3.4 Answer Refinement

In most cases, the set of candidate answers found by model Θ may already be a reasonable estimate of A_Q. Sometimes, however, noise may exist in the candidate set and has a higher probability than true answers. One possible reason is that relations in KG exhibit various types of patterns and properties. For example, relations like `FatherOf` and `liveIn` are asymmetric, but relations like `IsFriendWith` are symmetric property. Ideally, a good model should be able to learn all combinations of different properties, like symmetry, asymmetry, and transitivity. However, due to the incompleteness and complexity of the knowledge graph, it is almost impossible to find a perfect model which can satisfy all the properties and find answers without errors. If we could refine the results, the accuracy of the model may be further increased.

Based on this observation, we propose an answer refinement model to re-order the *top-k* candidates of the answer scoring module. Given the *top-k* candidates which have the highest scores found by the answer scoring module, we want to re-select the high quality entities from them. Inspired by [59, 149], given a topic entity v_Q and a path P, we concatenate them with each of the candidates to get k sequences. We treat each sequence as a language sentence and pass it through a Transformer [149] to predict the soundness of this sentence.

$$h_i = \text{TRANSFORMER}([\mathbf{e}_{v_Q}|\mathbf{r}_1|\ldots|\mathbf{r}_n|\mathbf{e}_{v_i}]) \tag{5.19}$$

where h_i is the output embedding of entity v_i. The final score is predicted by passing h_i through a feed forward neural network with Sigmoid function:

$$Pr(v_i|P, v_Q, \mathcal{G}) = \text{Sigmoid}(\text{FFN}(h_i)) \tag{5.20}$$

In the last step, a classifier will be used to return the answer predicted by the answer scoring module or the transformer module. The architecture is shown in Fig. 5.3. Note that the Refinement step is optional.

5.3.5 Learning Algorithm

The overall loss function of the proposed BiNet is as follows:

$$\mathcal{L} = \mathcal{L}_{KGQA} + \mathcal{L}_{KGC} + \mathcal{L}_{Path} + \mathcal{L}_{REG} \tag{5.21}$$

$$= \sum_{Q\in\overline{Q}} \mathcal{J}(\hat{y}, y) + \lambda_1 \sum_{(h,r,t)\in\mathcal{G}} \mathcal{J}(\hat{t}, t) + \lambda_2 \sum_{Q\in\overline{Q}} \mathcal{L}(\hat{P}, P) + \lambda_3||\mathbf{W}||_2^2 \tag{5.22}$$

where λ_1, λ_2 and λ_3 are hyper parameters used to balance the loss. $\overline{Q}$ is the KGQA training set and (h, r, t) belongs to KGC training set. $\mathcal{L}_{KGQA}$ is the KGQA loss, and $\mathcal{L}_{KGC}$ is the KGC loss. Both of them are calculated by $\mathcal{J}$ which is binary cross entropy loss. The first term

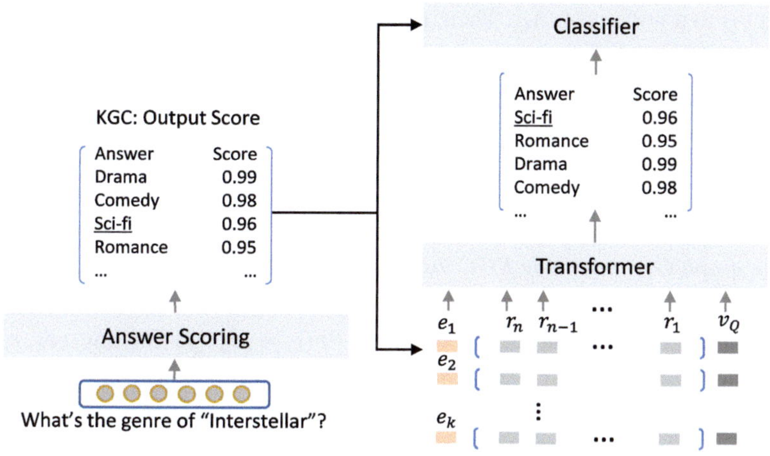

Fig. 5.3 Answer refinement

$\mathcal{J}(\hat{y}, y)$ measures the loss of the KGQA task, $\hat{y}$ is the answer predicted by BiNet and y is the ground truth. The second term $\mathcal{J}(\hat{t}, t)$ calculates the loss in the KGC task, $\hat{t}$ is the answer predicted by BiNet and t is the ground truth. The third item $||\mathbf{W}||_2^2$ is the regularization term for preventing overfitting. The last term $\mathcal{L}(\hat{P}, P)$ is the path decoder loss. Note that the KGQA task $\mathcal{L}_{KGQA}$ contains two parts. The first part is the loss of answer scoring model before answer refinement. The second part is the loss of answer scoring model after answer refinement.

5.3.6 Proof and Analysis

In this section, we analyze the expressive power of our model. In particular, we show in Lemmas 5.1 and 5.2 that KGC and KGQA can indeed mutually benefit each other. We further show in Lemma 5.3 that if the exact matching of a path exists in the knowledge graph and Eq. (5.16) is equal to 1 for any $(h, r, t) \in \mathcal{L}$, the proposed BiNet is guaranteed to find it.

Lemma 5.1 *(Benefit of KGQA for KGC) Given a language question $Q = (w_1, w_2, \ldots, w_{|Q|})$ with topic entity v_Q and answer set A_Q. Path $P = (r_1, r_2, \ldots, r_{|P|})$ is a relation sequence of Q. If there is no exact matching between v_Q and any $v \in A_Q$ in the knowledge graph $\mathcal{G}$, suppose $< v_k, r_j, v_m >$ is the missing link in the knowledge graph which is on the path from v_Q to a node $v_a \in A_Q$, minimizing KGQA loss $\mathcal{L}_{KGQA}$ will also minimize KGC loss $\mathcal{L}_{KGC}$.*

Proof Given an multi-hop path $P = (r_1, r_2, \ldots, r_{|P|})$, minimizing $\mathcal{L}_{KGQA}$ means maximizing $Pr(v|P, v_Q, \mathcal{G})$ for any $v \in A_Q$. Note that $< v_k, r_j, v_m >$ is the missing link on the path from v_Q to a node $v_a \in A_Q$. Then, maximizing $Pr(v_a|P, v_Q, \mathcal{G})$ means maximizing $Pr(v_m|r_j, v_k, \mathcal{G})$ which is the output of the KGC task on triple (v_k, r_j, v_m). Therefore, we have that $\mathcal{L}_{KGC}$ is minimized. □

Lemma 5.2 (Benefit of KGC for KGQA) *Given a language question* $Q = (w_1, w_2, \ldots, w_{|Q|})$ *with topic entity* v_Q *and answer set* A_Q. *Path* $P = (r_1, r_2, \ldots, r_{|P|})$ *is a relation sequence of* Q. *If there is no exact matching between* v_Q *and any* $v \in A_Q$ *in the knowledge graph G, minimizing KGC loss* $\mathcal{L}_{KGC}$ *will also minimize KGQA loss* $\mathcal{L}_{KGQA}$.

Proof Minimizing $\mathcal{L}_{KGC}$ means maximizing $Pr(v_t|r_i, v_h, \mathcal{G})$ for any $(v_h, r_i, v_t) \in \mathcal{G}$. Given a path $P = (r_1, r_2, \ldots, r_{|P|})$ and a topic entity v_Q, for any $v_a \in A_Q$, the probability $Pr(v_a|P, v_Q, \mathcal{G})$ can be expressed as Eq. (5.14). Since each item is maximized, $Pr(v_a|P, v_Q, \mathcal{G})$ is also maximized. Therefore, $\mathcal{L}_{KGQA}$ is minimized. □

If the background knowledge graph is complete, traditional path traverse or subgraph matching methods could find exact matches from knowledge graph $\mathcal{G}$, where each exact match is a subgraph in $\mathcal{G}$ which could be mapped to multi-hop query $(v_Q, P, ?)$ exactly. Ideally, if the answer scoring model in BiNet could output 1 for any triple (h, r, t) in $\mathcal{G}$, BiNet could also find exact matches like subgraph matching methods.

Lemma 5.3 (Model Soundness) *Given a natural language question* $Q = (w_1, w_2, \ldots, w_{|Q|})$ *with topic entity* v_Q *and answer set* A_Q. *Path* $P = (r_1, r_2, \ldots, r_{|P|})$ *is a relation sequence of* Q. *If Eq. (5.16) is equal to 1 for any* $(h, r, t) \in \mathcal{L}$, *and if there is an exact match between* v_Q *and any* $v \in A_Q$ *in the knowledge graph* $\mathcal{G}$, *the proposed* BiNet *is guaranteed to find it as long as* $k \geq D^{|P|-1}$, *where D is the maximum out-degree of a node on a specific relation type in the knowledge graph.*

Proof Because D is the maximum out-degree of a node on a specific relation type in the knowledge graph, there are at most $D^{|P|-1}$ exact matching paths of $(r_1, r_2, \ldots, r_{|P|-1})$ exist in the knowledge graph started from topic entity v_Q. If $k \geq D^{|P|-1}$, they will be all included in the *top-k* candidates. Therefore, the exact match can be found. □

Note that $|P|$ is a relatively small value (e.g., usually $|P| \leq 3$).

5.4 Experiments

In this section, we evaluate the performance of the proposed BiNet on several public datasets. We first introduce the datasets and baselines used in the chapter, and then present the experiment results.

5.4.1 Experimental Setting

Three datasets are used in the chapter which are listed below:

- **MetaQA** is a multi-hop question dataset on movie domain which contains more than 400K natural language questions. The background knowledge graph contains more than 100K triples which includes different relationships among directors, movies, genres and actors. All questions can be divided into three categories: 1-hop, 2-hop and 3-hop.
- **WebQuestionsSP** contains about 4,000 questions which could be answered by Freebase. It is a mixture of 1-hop questions and 2-hop questions.
- **SimpleQuestions** [2] is a dataset which consists of more than 100K simple 1-hop natural language questions and their corresponding triples from Freebase. We use a subset of SimpleQuestions which contains all the questions that can be answered by Freebase used in WebQuestionsSP.

In the experiment, we compare the proposed BiNet with baselines in the challenging settings with incomplete KG with 50 and 70% missing edges (we randomly delete 50 and 70% edges from the full knowledge graph). This is because KGQA on the complete KG becomes trivial on these datasets. For example, simply using path traverse or subgraph matching could achieve nearly 100% accuracy.

We compare our method BiNet with 4 baselines on the KGQA task, including:

- GraftNet [136] finds a question-specific subgraph containing KG facts, and then uses a graph neural network to predict the answers.
- PullNet [135] utilizes the shortest path as supervision to train graph retrieval module and conduct multi-hop reasoning with GraftNet on the retrieved sub-graph.
- Key-Value Memory Network (KVMem) [103] maintains a memory table which stores KG facts and uses this for retrieval.
- EmbedKGQA [126] conducts multi-hop reasoning through matching pre-trained entity embeddings with question embedding obtained from RoBERTa. We use EmbedKGQA without relation matching.

We compare our method BiNet with 4 baselines on the KGC task, including

- RESCAL [107] is a three-way tensor factorization method which embeds each entity as a latent vector and each relation as a matrix.
- DistMult [162] uses low dimensional vectors to represent nodes and uses bilinear functions to represent relations.

[2] The query paths of WebQuestionsSP and SimpleQuestions are given, so we don't need to use Question Encoder-Decoder.

- ComplEx [147] embeds each entity as a complex vector which contains a real part and an imaginary part, and the relations are represented as bilinear functions.

5.4.2 Knowledge Graph Question Answering Performance

Following the standard setup in KGQA [126], we evaluate the accuracy using the Hits@1 metrics. Table 5.2 shows the results of all baseline methods on the MetaQA dataset with 50% incomplete KG and 30% incomplete KG, respectively. The code of PullNet is not publicly available, so we omit its performance on 30% incomplete knowledge graph. The results of GraftNet and KV-Mem are from [136]. As we can see from the table, when the background knowledge graph becomes sparse, the Hits@1 accuracy of all methods decreases. This means that the quality of the background knowledge graph has a significant impact on the KGQA task. For subgraph retrieval-based methods: GraftNet and PullNet, their performances suffer severely from the incompleteness of the background knowledge graph. This is because, when the KG becomes sparse, the generated subgraphs of these methods are unable to cover the answer entities. Among all the methods, our method achieves the best results. For 50% incomplete KG, our BiNet is about 2.5% better than EmbedKGQA and more than 25% better than all other baseline methods. For 30% incomplete KG, our BiNet is about 3.5% better than EmbedKGQA and 28% better than other baseline methods on average.

Table 5.3 shows the performance of different methods on WebQuestionsSP and SimpleQuestions datasets. We have the similar results. When the background knowledge graph becomes sparse, the accuracy of all methods decreases. Nonetheless, the proposed BiNet consistently outperforms other baseline methods, and it is about 1.2% better than baseline methods on average.

Table 5.2 KGQA Hits@1 results of MetaQA on 50 and 30% incomplete knowledge graphs

Model	50% KG				30% KG			
	MetaQA-1	MetaQA-2	MetaQA-3	Avg	MetaQA-1	MetaQA-2	MetaQA-3	Avg
GraftNet	64.0	52.6	59.2	58.6	48.4			48.4
PullNet	65.1	52.1	59.7	59.0	–	–	–	–
KV-Mem	63.6	41.8	37.6	47.7	44.7			44.7
EmbedKGQA	83.1	91.8	70.3	81.7	77.7	81.2	69.0	76.0
BiNet	*84.2*	*92.8*	*75.9*	*84.3*	*77.8*	*86.4*	*74.3*	*79.5*

Table 5.3 KGQA Hits@1 results of WQSP and SimpleQA on 50 and 30% incomplete knowledge graphs

Model	50% KG		30% KG	
	Webqsp	SimpleQA	Webqsp	SimpleQA
GraftNet	32.7	39.8	34.9	25.7
PullNet	48.2	–	34.6	–
KV-Mem	*50.1*	28.9	25.8	22.8
EmbedKGQA	47.3	41.7	38.8	33.5
BiNet	49.4	*42.6*	*40.5*	*33.9*

5.4.3 Knowledge Graph Completion Performance

Figure 5.4a, b show the accuracy of different knowledge graph completion methods on 50% incomplete knowledge graph and 30% incomplete knowledge graph, respectively. Traditional knowledge graph embedding methods like RESCAL, DistMult and ComplEx do not perform very well on MetaQA knowledge graph. This is because the knowledge graph is very sparse. It only contains about 66,791 edges which is 2.3% that of Freebase. Since EmbedKGQA is not designed for knowledge graph completion, when using EmbedKGQA for knowledge graph completion, we first transform the KG triple (h, r, v) to a natural language question, and then train EmbedKGQA. As we can see, for MetaQA knowledge graph, EmbedKGQA performs quite well. It is higher than other existing knowledge graph completion baseline methods without the question data. When using the question date, EmbedKGQA is about 15% higher than other baselines. Compare with other methods, the proposed BiNet consistently has the highest KGC performance.

5.4.4 Ablation Studies

In this section, we evaluate the effectiveness of each component of the proposed BiNet.

A—Answer Refinement. In this subsection, we show the effectiveness of the answer refinement module. The results are shown in Table 5.4. As we can see, the refinement module could improve the prediction accuracy by about 2% on average on both 50% incomplete and 30% incomplete knowledge graphs. This means that the proposed refinement model indeed alleviates the sparsity of the background knowledge graph. Compared with 1-hop natural language questions, e.g., Webqsp and SimpleQA, the accuracy improvement on long path questions is more significant (3.2% vs 1.3%). This means that when the path becomes longer, the refinement module is even more effective.

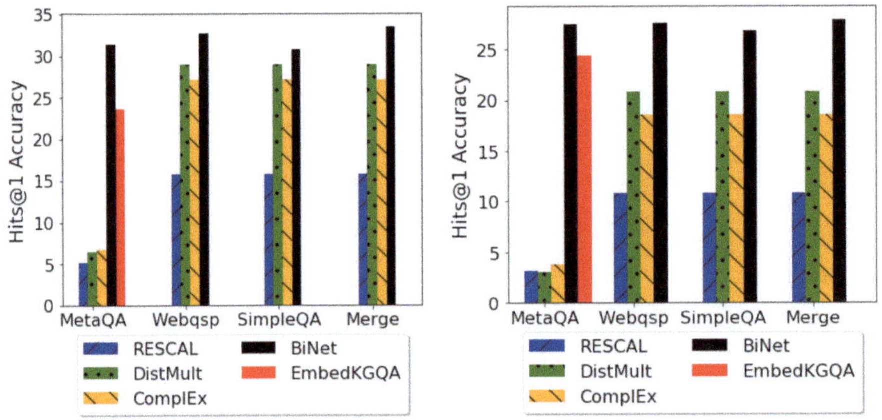

(a) Accuracy on 50% Incomplete KG (b) Accuracy on 30% Incomplete KG

Fig. 5.4 Knowledge graph completion accuracy. ‘Merge’ dataset is the combination of Webqsp and SimpleQA

Table 5.4 Ablation study of answer refinement

Model	MetaQA-3hop	Webqsp	SimpleQA
50% KG			
BiNet without re-ranking	70.3	47.2	41.8
BiNet with re-ranking	*75.9*	*49.4*	*42.6*
30% KG			
BiNet without re-ranking	71.2	39.1	33.2
BiNet with re-ranking	*74.3*	*40.5*	*33.9*

B—The Power of Knowledge Graph Completion. In this subsection, we study whether simply completing the knowledge graph first could help KGQA task. When completing the knowledge graph, we use two strategies. The first one is the heuristic based method to complete the knowledge graph according to the natural language questions in the KGQA training and validation sets. This heuristic based method is similar to some rule based knowledge graph completion methods [47]. For example, if `A` is the child of `B` and `C` is the husband of `B`, then `A` is the child of `C`. After we complete the knowledge graph by the heuristic method, we further use the second strategy to complete the knowledge graph. We train ComplEx on the partially completed knowledge graph. Then given a triple $(h, r, ?)$, we use ComplEx to predict the answer. Note that we only keep those triples which satisfy $Pr(v_t|r_i, v_h, \mathcal{G}) >= 0.99$ where 1 is the highest score. We also prune some obvious wrong predictions from the triple list. For example, if the head entity is a movie `Interstellar`

Table 5.5 The power of knowledge graph completion

Model	EmbedKGQA	KGC + EmbedKGQA	BINET
50% KG			
MetaQA-1hop	83.1	83.2	84.2
MetaQA-2hop	91.8	92.4	92.8
MetaQA-3hop	70.3	73.5	75.9
Webqsp	47.3	47.7	49.4
SimpleQA	41.7	41.9	42.6
30% KG			
MetaQA-1hop	77.7	77.8	77.8
MetaQA-2hop	81.2	85.1	86.4
MetaQA-3hop	69.0	71.1	74.3
Webqsp	38.8	39.1	40.5
SimpleQA	33.5	33.7	33.9

and the relation is `starredBy`, but the prediction is not a movie star, we delete this triple even though its probability is greater than or equal to 0.99.

Table 5.5 shows the performance of KGC+EmbedKGQA and the proposed BINET. The performance of KGC+EmbedKGQA is better than EmbedKGQA on all three kinds of questions. This means that completing the knowledge graph first can indeed improve the KGQA performance. In MetaQA-2hop questions on 30% incomplete KG, it achieves the largest accuracy improvement which is 3.9%. On average, completing the knowledge graph first could improve about 1.2% Hits@1 accuracy. The proposed BINET further improves the performance by 1.3%.

C—Path Prediction. A good path decoder is important for the proposed model BINET. Before training the path decoder, we manually add ground-truth paths to the training data. When training the model, at each time step the path decoder will predict what the next relation is according to the previous relation decoded. With probability equal to the α ($\alpha = 0.5$), the decoder will use the actual ground-truth relation as the input to the decoder during the next time-step. However, with probability $1 - \alpha$, it will use the relation that the model predicts as the next input to the model, even if it does not match the actual next relation in the ground-truth. Table 5.6 shows some results of the path decoder. As we can see, the path decoder could generate relation paths with high accuracy.

D—Efficiency. Figure 5.5 show the training time and test time of BINET on different datasets. As we can see, the runtime of BINET on Webqsp and SimpleQA is much larger than that on MetaQA dataset. This is because the background knowledge graph of Webqsp and SimpleQA

Table 5.6 Results of path decoder

Question	Path
The movies starred by [Tanner Maguire] were in which genres	Starred_actors_reverse I has_genre
When did the movies written by [Cristian Nemescu] release	Written_by_reverse I release_year
The films acted by [Benjamin Pitts] were released in which years	Starred_actors_reverse I release_year
Who are movie co-writers of [Ray Ashley]	Written_by_reverse I written_by
Who co-starred with [Mary McDonnell]	Starred_actors_reverse I starred_actors
Who are movie co-directors of [Jack Hazan]	Directed_by_reverse I directed_by

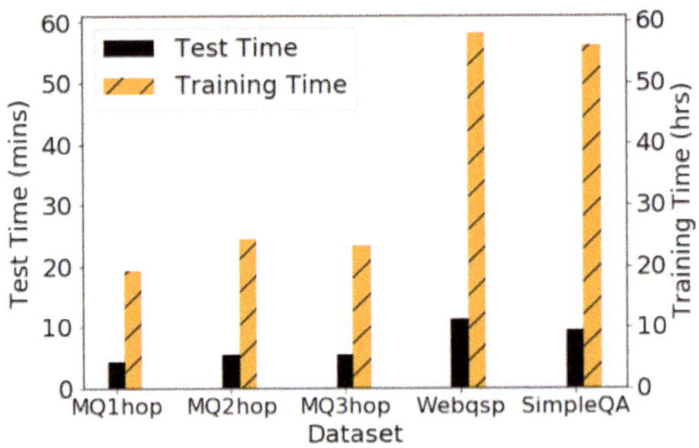

Fig. 5.5 BiNet training and test time

is much larger than that of MetaQA. Despite the long training time, the test time of BiNet is relatively short which is less than 15 min.

5.5 Discussion

Enhancing knowledge graph completion with symbolic path information in multi-hop question answering has yielded promising results in this study. Future integration of neural network-based models for reasoning within the embedding space could enhance both tasks, demonstrating the effectiveness of neural-symbolic reasoning. However, the current approach only utilizes explicit triplet information from the knowledge graph, missing out on crucial implicit information such as subgraph structures. Exploring how to effectively leverage these additional details represents a promising direction for future research.

Accurate Query Answering with LLMs Over Incomplete KG

6

Despite the superb performance in many tasks, large language models (LLMs) bear the risk of generating hallucination or even wrong answers when confronted with tasks that demand the accuracy of knowledge. The issue becomes even more noticeable when addressing logic queries that require multiple logic reasoning steps. On the other hand, knowledge graph (KG) based question answering methods are capable of accurately identifying the correct answers with the help of knowledge graph, yet its accuracy could quickly deteriorate when the knowledge graph itself is sparse and incomplete. It remains a critical challenge on how to integrate knowledge graph reasoning with LLMs in a mutually beneficial way so as to mitigate both the hallucination problem of LLMs as well as the incompleteness issue of knowledge graphs. In this chapter, we introduce 'Logic-Query-of-Thoughts' (LGOT) which combines LLMs with knowledge graph based logic query reasoning. LGOT seamlessly combines knowledge graph reasoning and LLMs, effectively breaking down complex logic queries into easy to answer subquestions. Through the utilization of both knowledge graph reasoning and LLMs, it successfully derives answers for each subquestion. By aggregating these results and selecting the highest quality candidate answers for each step, LGOT achieves accurate results to complex questions.

6.1 Introduction

Large language models (LLMs) have exhibited remarkable performance across various natural language processing (NLP) tasks, including question answering [12], machine translation [6], text generation [139], recommender system [39] and so on. Recent increases in the model size have further enhanced the learning capabilities of LLMs, propelling their per-

L. Liu and H. Tong, *Neural Symbolic Knowledge Graph Reasoning*, Synthesis Lectures on Computer Science, https://doi.org/10.1007/978-3-032-15858-1_6

formance to new heights. Notable examples of these advanced models include GPT-4 [109] and Llama 2 [146], among others.

Despite the great success in various domains, LLMs have faced criticism for their limited factual knowledge. Specifically, LLMs tend to memorize facts and knowledge present in their training data [112]. However, research has revealed that LLMs struggle with factual recall and could generate factually incorrect statements, leading to hallucinations [111]. For instance, when asked, '`When did Einstein discover gravity?`', LLMs might erroneously respond with '`1687`,' which contradicts the fact that Isaac Newton formulated the theory of gravity [111]. This issue significantly undermines the reliability of LLMs, especially in contexts requiring multi-step reasoning. Efforts have been made to address these limitations using approaches like Chain-of-Thought [154], but our experiments reveal that even these methods still fall short of mitigating these issues.

Different from Large Language Models (LLMs), knowledge graphs store structured human knowledge, making them a valuable resource for finding answers. Knowledge Graph Question Answering (KGQA) aims to identify an answer entity within the knowledge graph to respond to a given question. Compared with LLMs, KGQA generates more accurate results when the knowledge graph is complete. However, the performance of KGQA deteriorates quickly when the underlying KG itself is incomplete with missing relations. A natural question arises: *how can we combine knowledge graph question answering with LLMs so that they can mutually benefit each other?*

In this chapter, we introduce a novel model that combines Large Language Models (LLMs) with knowledge graph reasoning to enhance question-answering capabilities. Drawing inspiration from the concepts of Chain-of-Thoughts [154], Tree-of-Thoughts [164] and Graph-of-Thoughts [8], we present 'Logic-Query-of-Thoughts' (LGOT) which answers complex logic queries by reasoning on multiple subquestions. Different from CoT, ToT and GoT which generate thoughts automatically, LGOT incorporates a logical query structure and human-defined logic transformations to guide LLMs and knowledge graph reasoning algorithm in finding answers. More precisely, LGOT utilizes a combination of knowledge graph reasoning methods and LLMs to identify potentially correct answers for each subquestion. These results are merged to generate a comprehensive answer set. Then, the merged answers serve as input for subsequent subquestions. This iterative process continues until the final answers are obtained. Our experimental results demonstrate that the proposed LGOT significantly enhances the performance, with up to 20% improvement over ChatGPT.

6.2 Problem Definition

A knowledge graph can be denoted as $\mathcal{G} = (\mathcal{E}, \mathcal{R}, \mathcal{L})$ where $\mathcal{E} = \{e_1, e_2, \ldots, e_n\}$ is the set of nodes/entities, $\mathcal{R} = \{r_1, r_2, \ldots, r_m\}$ is the set of relations and $\mathcal{L}$ is the list of triples. Each triple in the knowledge graph can be denoted as (h, r, t) where $h \in \mathcal{E}$ is the head (i.e., subject) of the triple, $t \in \mathcal{E}$ is the tail (i.e., object) of the triple and $r \in \mathcal{R}$ is the edge (i.e., relation, predicate) of the triple which connects the head h to the tail t.

Complex logical queries, or Existential First-Order (EFO) queries studied in most literature [153, 166], can be represented with First-Order Logic (FOL) formula that includes conjunction ($\wedge$), disjunction ($\vee$), negation ($\neg$), and the existential quantifier ($\exists$). Let $v_1, \ldots, v_k$ be existentially quantified variables, and $v_?$ be the target variable. An EFO query is written as

$$q[v_?] = v_?.\exists v_1, \ldots, v_k : \pi_1 \vee \pi_2 \vee \ldots \vee \pi_n, \tag{6.1}$$

where $\pi_i = \varrho_1^i \wedge \varrho_2^i \wedge \ldots \wedge \varrho_m^i$ is a conjunctive query. Each ϱ_m^i denotes a logical atom formula or its negation on relation r between an entity and a variable or two variables. That is, it can be one of $r(u, v)$, $\neg r(u, v)$, $r(v_1, v_2)$ or $\neg r(v_1, v_2)$, where $u \in \mathcal{E}$ is an entity.

When answering complex logical queries, we can directly utilize language models to answer them, or in many cases [166], we can transform the complex logical query as a computation graph [117] with operator nodes of projection, intersection, negation, and union. This directed graph demonstrates the computation process to answer the query. Each node of the computation graph represents a distribution over a set of entities in the KG, and each edge represents a logical transformation of this distribution. An example is given in Figure 6.1. The mapping along each edge applies a certain logical operator:

Projection operation. Given an input set $S_{\text{in}} \subseteq \mathcal{E}$ and relation $r \in \mathcal{R}$, the projection operation obtains $S_{\text{out}} = \cup_{e \in S} Ar(e)$, where $Ar(e) \equiv \{e_i \in \mathcal{E} : r(e, e_i) = \text{True}\}$.

Intersection operation. Given two input sets $S_{\text{in},1}$ and $S_{\text{in},2}$, the intersection operation outputs $S_{\text{out}} = S_{\text{in},1} \cap S_{\text{in},2}$.

Negation operation. Given a set of entities $S \subseteq \mathcal{E}$, the negation operation computes its complement $\overline{S} \equiv \mathcal{E} \setminus S$.

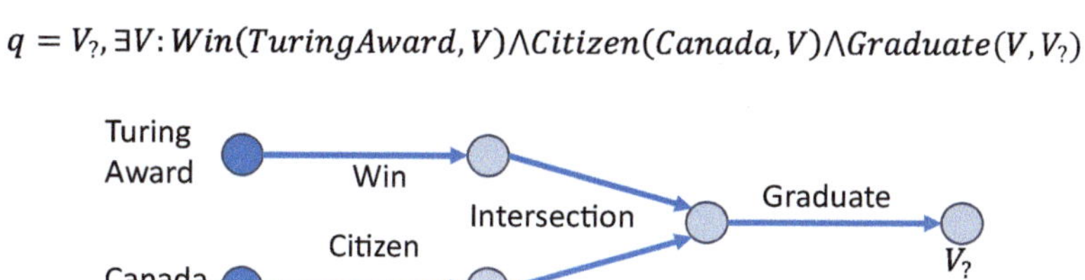

Fig. 6.1 An example of a logic query of question 'Where did Canadian citizens with Turing Award graduate?' [117]

Union operation. Given a set of entities $S_1 \subseteq \mathcal{E}$ and $S_2 \subseteq \mathcal{E}$, the union operation computes their union $S \equiv S_1 \cup S_2$.

6.2.1 Fuzzy Logic

Different from Boolean logic, fuzzy logic assigns a value in [0, 1] to every logical formula. A t-norm $\top : [0, 1] \times [0, 1] \rightarrow [0, 1]$ represents generalized conjunction in fuzzy logic. Analogously, t-conorms are dual to t-norms for disjunction in fuzzy logic, the t-conorm is defined as $\perp(x, y) = 1 - \top(1 - x, 1 - y)$. The negator is $n(x) = 1 - x$. When applying fuzzy logic to FOL queries, following [7, 23, 104], the complex logical query answering problem can be formulated as

$$q[v?] = \arg \max_{v_1,\ldots,v_N \in \mathcal{E}} ((\varrho_1^1 \top \ldots \top \varrho_{m_1}^1) \perp \ldots \perp (\varrho_1^n \top \ldots \top \varrho_{m_n}^n)) \tag{6.2}$$

where $\varrho_j^i \in [0, 1]$ is scored by a pretrained KGE model according to the likelihood of the atomic formula. In this chapter, we use the product t-norm, where given $a, b \in [0, 1]$: $\top(a, b) = a \cdot b, \quad \perp(a, b) = 1 - (1 - a)(1 - b)$.

6.2.2 Problem Definition

When answering complex logical queries, finding the answers directly in the knowledge graph is hindered by its incompleteness. On the other hand, using LLMs to answer complex logical queries may result in incorrect answers due to the hallucination problem. This chapter aims to study the integration of LLMs and fuzzy logic reasoning to effectively answer logical queries by leveraging the advantages of both approaches. Formally, the problem is defined as follows.

Given: (1) A knowledge graph $\mathcal{G}$, (2) a complex logical query, (3) a large language model, (4) a fuzzy logic reasoning model; **Output:** (1) the answer of the question.

6.3 Proposed Method

In this chapter, we design a model, LGOT, to answer complex logical queries by combining LLM and KGQA (We use KGQA to denote finding answers in the knowledge graph through fuzzy logic reasoning), in a mutually beneficial way to leverage the advantages of both. Because large language models are not good at answering complex logical queries, LGOT decomposes the original complex logical query into multiple easy-to-answer subquestions according to the atom predicates in its dependency graph and executes each subquestion in their corresponding order in the computation graph. For each atom predicate, both the

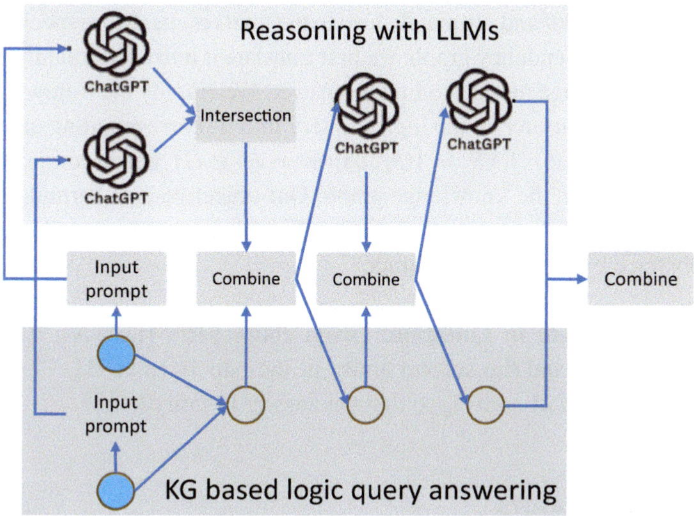

Fig. 6.2 Framework of the proposed LQOT

large language model and fuzzy logic reasoning method are used to solve it according to its corresponding logical operator. Figure 6.2 introduces the framework of LGOT, where the same operator is executed by both the LLMs and KGQA, and LGOT combines the input and output.

Our proposed method has several advantages compared with existing methods. Previous retrieval-augmented generation-based methods, e.g., [28], usually fail to find the correct answers due to the incompleteness of the knowledge graph. Therefore, we let the language model call the fuzzy logic reasoning method and use its answers to augment the LLM's answers. At the same time, the answers generated by the LLMs for each atomic question are also used to augment the results of the fuzzy logic reasoning method in case the knowledge graph is incomplete. This ensures that these two models can mutually assist each other.

We chose fuzzy logic reasoning due to several reasons, first, fuzzy vector can handle different types of logical operations, and all these operations are closed. Second, fuzzy vector can effectively connect LLMs with fuzzy logic reasoning model since we only need to update the value of the fuzzy vector according to the results of LLMs, which is much more convenient compared to relearning the box embedding [85, 117] or probability distribution [118]. The details of the method are introduced in the following sections.

6.3.1 Question Transformation for LLMs

Answering complex logical queries requires multiple reasoning steps. To address this, we divide the original hard-to-answer question into multiple easy-to-answer subquestions, which are executed in order and ultimately lead to the final results. More specifically, for each atom predicate in the dependency graph, we first translate it into a natural language question and then use large language models to find its answer. We employ the following approach to transform the atomic relation r in the logical query into its corresponding question prompt.

Given an atom predicate r, let $S_r = \{(h_i, t_i)|(h_i, r, t_i) \in G\}$, which comprises of all entity pairs associated with r in the knowledge graph. Our objective is to formulate a query that facilitates the retrieval of t_i given both h_i and r. For this purpose, we utilize the following prompt to transform r to a question prompt.

Prompt: Atom Predicate to Question: Given entity pairs {text_A}. The relationship between the first entity and the second entity in the pair is {text_B}. Please rewrite the relationship to a question of entity1, so that the answer is entity2.

6.3.2 Logical Operators

So far, we have described how to transform each atom predicate in the dependency graph into a natural language question. Now, we will describe how to use both large language models and fuzzy logic reasoning models to find answers. To answer a query according to the computation graph, we need to define logical operators for both the language models and the fuzzy logic reasoning method. In the following section, we describe the design of the different logical operators used in the computation graph.

Projection operation: given an entity set S_1, the goal of projection operation is to find answer set S_2 which is connected to any entities in S_1 with relation r in any case.

For large language models, the projection operation takes a set of entities and a question prompt as input and generates an additional set of answers. To achieve this with LLMs, we employ the following prompt construction method: given a set of input entities S, we concatenate them using the word "`or`". For example, if $S = \{\texttt{TomHanks}, \texttt{TomCruise}\}$, the resulting prompt would be '`Tom Hanks or Tom Cruise.`' Each atom predicate corresponds to a question prompt defined above, and we construct the input prompt by combining the question prompt with the relevant entities. Here is an example: Suppose we have the relation $(h, \texttt{isDirectedBy}, ?)$ with its question prompt '`which movies are directed by h?`' In this case, the input prompt would be '`which movies are directed by Tom Hanks or Tom Cruise?`'.

For fuzzy logic reasoning, we represent each variable in the computation graph as a fuzzy vector $T \in \mathbb{R}^{|\mathcal{E}|}$ [104], where the i-th element in the fuzzy vector indicates the probability that entity e_i is the correct answer for the current variable node. We define an operator $M_r \in [0, 1]^{|\mathcal{E}|\times|\mathcal{E}|}$ for each relation r, where $M_r(i, j)$ denotes the probability that e_i is connected with e_j by relation r. We utilize knowledge graph embedding methods $g_{r_i}(e_i, e_j)$ to score the likelihood that e_i is connected with e_j by relation r_i, where

$g()$ is calculated based on the embeddings of e_i and e_j. Many KGE methods can be used as $g()$, in this chapter, we adopt ComplEx [147] which is widely used in many applications. Note that because the knowledge graph is very sparse, M_r can be stored in a single GPU, and the reasoning process is efficient. Let $\max_i$ and $\max_j$ denote the maximum over rows and columns of a matrix, respectively. Then the projection operation for $(v_k, r, v_?)$ would be $T(v?) = \max_j((T(v_k)^\top \cdots \times |\mathcal{E}|) \odot M_r)$. If v_k is a constant entity $c = e_i$, then $T(v?) = \text{row}_i(M_r)$.

Intersection operation: give the answer sets of different subquestions, the intersection operator is used to get the answers that belongs to all question. For large language models, the intersection operation can be simply modeled as the intersection of two answer sets. When reasoning with fuzzy vector, if a node $v?$ is formed by merging child nodes $\{v_1, \ldots, v_K\}$ through intersection, then the intersection can be calculated by: $T(v?) = \prod_{1\leq i\leq K} T(v_i)$.

Negation operation: given a set of entities, the negation contains all entities that are not belong to this set. The negation for large language model is all entities in the knowledge graph which do not belong to the answer of LLMs. For fuzzy logic reasoning, the negation operation for $\neg r(v_k, v_?)$ would be $T(v?) = \max_j((T(v_k)^\top \cdots \times |\mathcal{E}|) \odot (\mathbf{I} - M_r))$. If v_k is a constant entity $c = e_i$, then $T(v?) = \text{row}_i(\mathbf{I} - M_r)$.

Union operation: The union operation of LLMs is the union of their answers. For fuzzy logic reasoning, if a node $v?$ is formed by merging child nodes $\{v_1, \ldots, v_K\}$ through union, then the union can be calculated by: $T(v?) = \mathbf{I} - \prod_{1\leq i\leq K}(\mathbf{I} - T(v_i))$.

6.3.3 Combination of LLMs and Fuzzy Logic Reasoning

When modeling the operations with LLMs, we can translate the input information into input prompts and get the output by parsing the LLM outputs. Given that the LLMs are generative models, the output set is assumed to be a very limited set of entities with estimated confidence values: $S_{\text{out}}^{\text{LLM}} = \{(e, p(e))\}$, where $|S_{\text{out}}^{\text{LLM}}| = K \ll |\mathcal{E}|$, $p(e)$ is the confidence for the entity e. We note that estimating $p(e)$ is non-trivial, and $p(e)$ can be zero in most cases (e.g., LLMs fail to output all the correct answers), so how to estimate $p(e)$ is an important task.

On the other hand, current logical query answering methods primarily employ retrieval-based approaches. In these methods, scores are assigned to all entities in $\mathcal{E}$ to indicate the likelihood that each entity is the correct answer. $S_{\text{out}}^{\text{KGQA}} = \{(e, \text{Score}(e)) : e \in \mathcal{E}\}$, where $|S_{\text{out}}^{\text{KGQA}}| = |\mathcal{E}|$. KGQA methods rank all existing entities based on their scores, which we assume to be within the range of 0 to 1 for convenience.

The success of LGOT heavily relies on combining the advantages of LLMs and fuzzy logic reasoning to mitigate their disadvantages. For LLMs, the outputs are featured with high accuracy of a small set of predictions but might suffer from randomness and even hallucination. For fuzzy logic reasoning, the outputs may suffer from knowledge graph incompleteness. We discuss how to obtain the final answer by obtaining the outputs $S_{\text{out}}^{\text{KGQA}}$ and $S_{\text{out}}^{\text{LLM}}$.

Likelihood ratio test for LLM outputs. Inspired by the likelihood ratio test [65], we use a likelihood ratio-based scheme to select potential correct answers from LLMs. Let $p(e_1), \ldots, p(e_N)$ denote the predicted likelihoods of the answers, and let $p_{\max} := \max_{i=1}^{N} p(e_i)$. Ideally, if there are K correct answers, a perfect LLM should predict $p(e_i) \approx \frac{1}{K} \approx p_{\max}$ for the correct answers, and $p(e_i) \approx 0$ for the wrong answers. Since K is unknown to us, we cannot directly decide their correctness via $p(e_i)$. Instead, we decide the correctness of each answer e_i through the *likelihood ratio* $\frac{p(e_i)}{p_{\max}}$. For each answer e_i, the likelihood ratio should be

$$\frac{p(e_i)}{p_{\max}} \approx \begin{cases} 1, & \text{if } e_i \text{ is correct;} \\ 0, & \text{if } e_i \text{ is wrong.} \end{cases} \tag{6.3}$$

Since the likelihood ratio $\frac{p(e_i)}{p_{\max}}$ does not depend on K, it allows us to decide their correctness without knowing the number of correct answers.

Enhancing KGQA with LLM. Since knowledge graphs are often incomplete, KGQA methods may encounter issues due to this incompleteness. To address this problem, we select answers with scores $p(e_i)/p_{\max} \geq \theta$ identified by LLMs. We then assign the score $\alpha * p(e_i)$ to the fuzzy vector obtained by the fuzzy logic reasoning method, thereby completing the fuzzy vector.

Enhancing LLM with KGQA. Answers generated by LLMs may be incorrect due to the hallucination problem. However, answers from fuzzy logic reasoning are more likely to be correct if they have a high score Score(). Therefore, after updating the fuzzy vector, we select the top-min$\{k, 10\}$ answers based on their scores and consider them as the correct answers to refine LLMs' output.

Answer Evaluation

After obtaining answers by combining LLMs with KGQA, we can further utilize an answer evaluator to refine the selection of answers. The purpose of the answer evaluator is to leverage Large Language Models (LLMs) to assess the quality of generated responses. Previous research has shown that LLMs are proficient at selecting the most appropriate answers from multiple choices, rather than directly generating answers [121].

For a given vertex v in the logic query, whether it represents an intermediate or final answer, answer evaluation is employed to gauge the likelihood that an entity e_i is the correct answer for v. More formally, when presented with a set of entities $S \subseteq \mathcal{E}$, the answer evaluation process selects a subset of entities from S that are likely to be correct answers for v. The prompt for the evaluation process is shown below. It is important to note that this module is optional.

Prompt: Answer Evaluation

Replace the answers with correct answers if the answers in the choices are not correct. Given the question {text_A} and its potential answer choices {text_B}, output top10 answers in a json format. The output json has key "input", "answer". The answers don't need to be in the potential answer choices.

An example of prompts used in LGOT can be found in Fig. 6.3.

6.4 Experiment

In this section, we evaluate the performance of the proposed LGOT on several public datasets. We first introduce the datasets and baselines used in the chapter, and then present the experiment results.

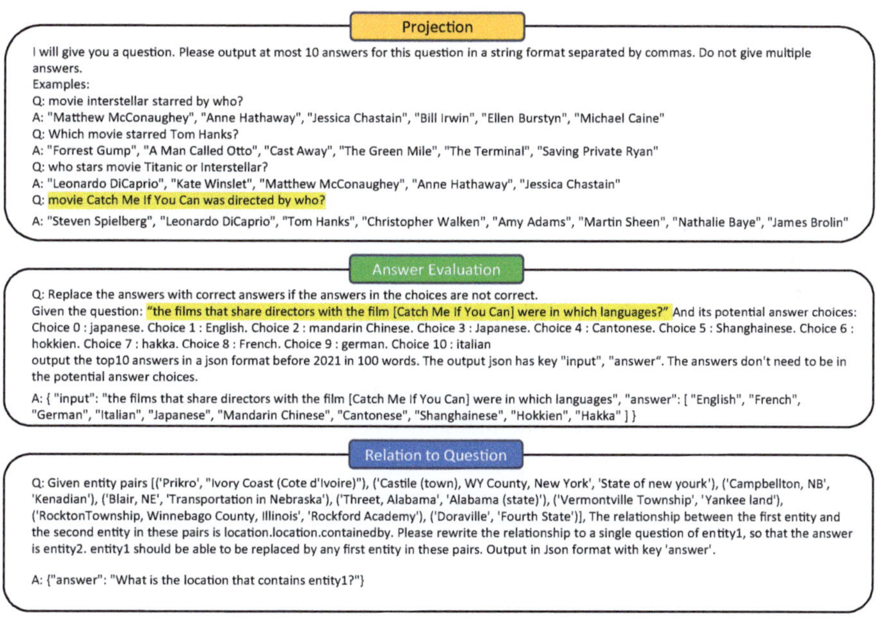

Fig. 6.3 Prompt examples

6.4.1 Experimental Setting

Existing logical query reasoning methods commonly test on dataset FB15k, FB15k-237, NELL995 and so on. One problem for these datasets is that their nodes and edges are represented as meaningless id which contains no semantic meaning. This seriously hinder the ability of large language models. To solve this problem, we use three different datasets: (1) **MetaQA** [126] is a multi-hop question dataset in the movie domain. The original MetaQA dataset only contains three types of questions, which are 1-hop, 2-hop and 3-hop questions. To support negation and union, we create new queries 2u and pin based on existing queries. (2) **ComplexWebQuestions** [70] contains the 1-hop, 2-hop and 3-hop questions. We only use the 3-hop questions in the experiment. (3) **GraphQuestions** [132] is a QA dataset consisting of a set of factoid questions with logical forms and ground-truth answers. We use 2-hop questions and 2 intersection questions in the experiment. All questions in ComplexWebQuestions and GraphQuestions datasets can be answered with Freebase. In the experiment, we compare the proposed LGOT with baselines on an incomplete KG with 50% missing edges (we randomly delete 50% of the edges from the full knowledge graph). When the knowledge graph is complete, traversing the graph alone can achieve 100% accuracy. **Note that all these datasets are complex logical queries**.

Six baselines are used in the experiments, including (1) ChatGPT [12]. (2) Chain-of-Thought [154]. (3) Query2Box [117]: a recent knowledge representation model that encodes logical queries as box embeddings. (4) BetaE [118]: a knowledge representation model, which utilizes the beta distribution to encapsulate logical queries. (5) QTO [7]: the state-of-the-art complex logical query answering method.

We adopt Hit ratio at k (Hit@k) as the metric to measure the performance of different baselines. Hit ratio at k (Hit@k) is the fraction of times a correct answer was retrieved within the top-k positions. We use $k = 1, 3$, and 10 in the experiments (Fig. 6.4).

6.4.2 Performance of Question Answering

Table 6.1 displays the question-answering performance of all models on the MetaQA dataset. Notably, LGOT surpasses all baseline methods when operating with the 50% incomplete knowledge graph. It demonstrates significant results with a 16% improvement in Hit@1, 12.8% in Hit@3, and 14.2% in Hit@10 compared to ChatGPT. Additionally, when compared to the logic query reasoning method QTO, our proposed approach achieves an even more

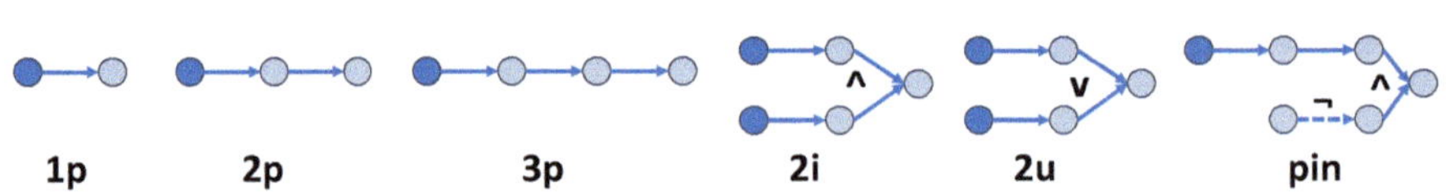

Fig. 6.4 Query structures, illustrated in their query computation tree representations

Table 6.1 Question answering results of MetaQA on 50% incomplete knowledge graphs

Dataset	1p			2p			3p			2u			pin		
Metric	Hit@1	Hit@3	Hit@10	Hit@1	Hit@3	Hit@10	Hit@1	Hit@3	Hit@10	Hit@1	Hit@3	Hit@10	Hit@1	Hit@3	Hit@10
Query2Box	19.7	57.5	72.7	14.1	44.7	70.1	20.1	33.7	47.7	70.4	75.3	77.8	36.1	40.5	45.5
BetaE	10.2	16.6	24.8	10.9	16.4	23.6	27.8	34.1	44.9	72.8	74.4	82.3	33.1	39.5	44.5
QTO	61.4	62.8	65.7	40.9	44.6	49.4	44.9	50.5	57.9	86.4	91.8	93.3	48.1	56.5	57.5
CoT	64.1	69.0	69.4	35.7	44.5	45.2	40.1	45.9	46.2	42.4	71.3	82.8	10.7	18.5	22.7
ChatGPT	66.1	80.3	84.3	45.4	57.0	64.1	41.6	52.6	55.6	44.0	77.9	88.8	11.2	21.1	29.8
LGOT	*77.7*	*86.3*	*89.7*	*58.7*	*79.0*	*83.7*	*67.9*	*83.9*	*90.8*	*90.4*	*98.3*	*99.4*	*52.6*	*62.8*	*68.7*

performance boost, with an average improvement of 17.5% in Hit@1, 24.4% in Hit@3, and 26.4% in Hit@10. We omit the results over complete knowledge graph, because both QTO and LGOT achieve near 100% accuracy.

One interesting observation is that ChatGPT does not perform well on 1-hop questions. One would expect ChatGPT to excel in answering short questions. However, we have found that ChatGPT is not proficient in responding to uncommon questions. For instance, when asked, "`What does [Walter Steiner] act in?`" the answers provided by ChatGPT include '`Downhill Racer (1969)`', '`The Great White Hope (1970)`', '`The Other Side of the Mountain (1975)`', '`The Other Side of the Mountain Part 2 (1978)`' and so on. However, none of these answers is correct. This illustrates ChatGPT's limitations in handling less common questions.

We also test the baselines performance on the other two datasets which are GraphQuestions and ComplexWebQuestions. Table 6.2 shows the results. As we can see, the proposed LGOT consistently outperforms all baseline methods with a large margin. For example, it can achieve 23.4% gain in Hit@1, 22.5% gain in Hit@3 and 19.5% gain in Hit@10 on GraphQuestions compared with ChatGPT in the 50% incomplete knowledge graph. For ComplexWebQuestions, LGOT can achieve 19.6% gain in Hit@1, 20.0% gain in Hit@3 and 24.7% gain in Hit@10 compared with ChatGPT in the 50% incomplete knowledge graph.

6.4.3 Ablation Study

A—ChatGPT + QTO. Figures 6.5, 6.6, and 6.7 illustrates the performance when simply combining LLMs with QTO. To elaborate, when given a logic query, we initially use QTO to obtain results, followed by employing ChatGPT to find additional results. These two sets of results are then combined. The results obtained by QTO, assigned with 100% probability, are placed at the beginning of the answer list, followed by the results from ChatGPT at the end. The results indicate that LGOT achieves an average improvement of 3.26% in MetaQA 2-hop questions and 6.1% in MetaQA 3-hop questions. For GraphQuestions, there is a 4.5% improvement in 2-hop questions and a 6.5% improvement in ComplexWebQuestions 3-hop questions. These results suggest that as the complexity of logic queries increases (i.e., requiring more reasoning steps), our proposed LGOT consistently outperforms the baseline methods.

B—The Effectiveness of Answer Evaluation. We compare two versions of LGOT: one with Answer Evaluation and one without. The results are displayed in Table 6.3. We employ Answer Evaluation in the ComplexWebQuestions 3-hop dataset, while for other datasets, we do not employ Answer Evaluation. As demonstrated in the table, the inclusion of Answer Evaluation leads to an average performance improvements of 6.2% for ComplexWebQuestions 3-hop questions. These results show the effectiveness of the proposed Answer Evaluation method.

Table 6.2 Question answering results of GraphQuestions and CWQ on full and 50% incomplete knowledge graphs

Dataset	GraphQuestions full KG			GraphQuestions 50% KG			CWQ 3hop full KG			CWQ 3hop 50% KG		
Metric	Hit@1	Hit@3	Hit@10	Hit@1	Hit@3	Hit@10	Hit@1	Hit@3	Hit@10	Hit@1	Hit@3	Hit@10
QTO	100.0	100.0	100.0	26.8	26.8	26.8	96.5	100.0	100.0	30.1	35.1	38.8
CoT	18.7	19.4	19.4	18.7	19.4	19.4	13.8	20.6	37.9	13.8	20.6	37.9
ChatGPT	18.4	19.0	21.5	28.8	30.3	33.3	13.7	20.7	34.5	13.7	20.7	34.5
LGOT	*100.0*	*100.0*	*100.0*	*52.2*	*52.8*	*52.8*	*96.5*	*96.5*	*100.0*	*33.3*	*40.7*	*59.2*

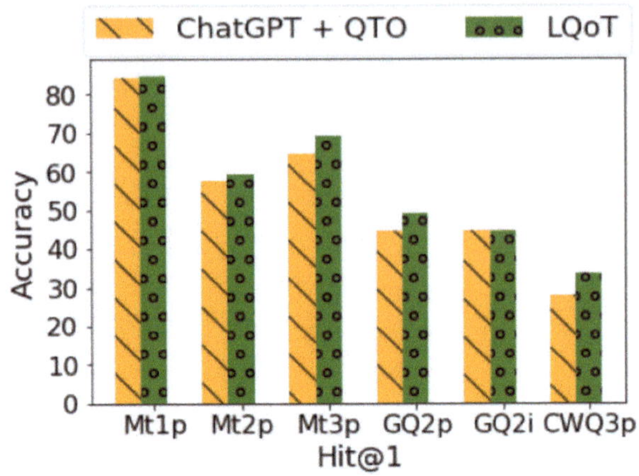

Fig. 6.5 Hit@1 performance

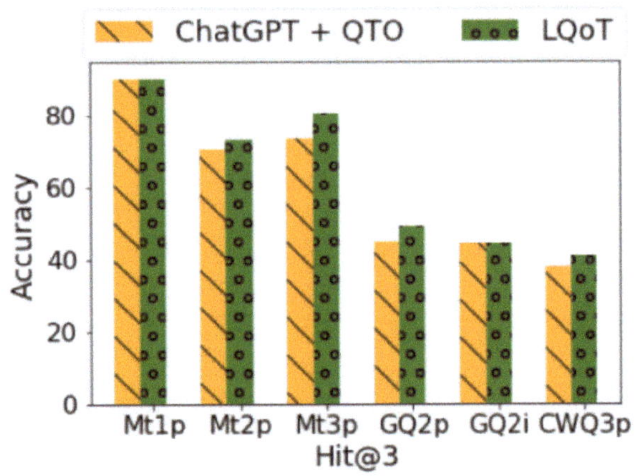

Fig. 6.6 Hit@3 performance

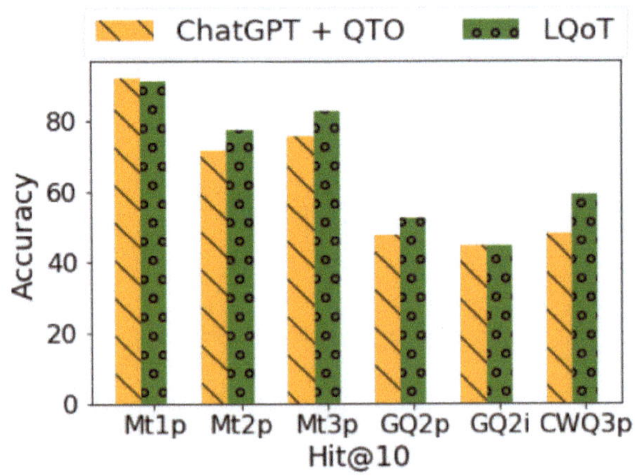

Fig. 6.7 Hit@10 performance

Table 6.3 Ablation study of answer evaluation

Model	LGOT no evaluation			LGOT		
	50% KG					
Metric	Hit@1	Hit@3	Hit@10	Hit@1	Hit@3	Hit@10
CWQ 3hop	29.6	33.3	51.8	*33.3*	*40.7*	*59.2*

6.5 Discussion

In this work, we focus on using large language models to address the issue of incomplete knowledge graph reasoning. While large language models contain a lot of information, they are prone to hallucinations when answering multi-step questions. On the other hand, simply relying on knowledge graph reasoning suffers from the incompleteness of the knowledge graph. By combining these two approaches, we can have them mutually support each other. However, the current integration of LLMs and knowledge graph reasoning is still shallow. In the future, we want to explore how to deepen the integration. Additionally, we aim to study what happens if the information from the large language model and the knowledge graph contradicts each other.

7 Ambiguous Query Answering with Neural Symbolic Reasoning Over Incomplete KG

Ambiguous Question Answering over Knowledge Graph refers to the task of accurately retrieving and interpreting answers from a knowledge graph in response to queries that are unclear or have multiple potential meanings. This process involves addressing the challenges posed by ambiguous language, which can lead to uncertainty in the interpretation of query intent. To overcome these challenges, advanced techniques such as semantic analysis, context understanding, and query reformulation are employed. By combining symbolic structural information from the knowledge graph with neural network models, we aim to clarify queries and find the most relevant and accurate answers. This chapter explores the application of neural-symbolic reasoning methods to resolve ambiguous questions effectively.

7.1 Introduction

Knowledge graph question answering (short for KGQA) aims to find entities in the knowledge graph which can correctly answer the given question. This problem has attracted great attention from both academia and industry, and abundant algorithms have been proposed recently. For example, RnG-KBQA [165] adopts a rank-and-generate approach to transform the input natural language question to a query graph and finds the answer according to the query graph. This strategy of finding answers based on the query graph can also be referred to as subgraph matching [81]. Key-Value Memory Network (KVMem) [103] stores KG facts in a memory table and uses it to retrieve entities which have the most similar embedding

L. Liu and H. Tong, *Neural Symbolic Knowledge Graph Reasoning*, Synthesis Lectures on Computer Science, https://doi.org/10.1007/978-3-032-15858-1_7

as the input query. EmbedKGQA [126] embeds both the input question and entities in the knowledge graph to points in the embedding space and finds answers according to their embedding similarity.

Despite the great progress, most works focus on answering *defectless* queries on knowledge graphs. These queries are assumed to be perfect and can precisely express users' query intentions. However, this is not true most of the time in real cases for the following reasons. First, the vocabulary of different users can vary dramatically. According to a prominent study on the human vocabulary problem [46], about 80–90% of the times two persons will give different representations when they are asked to name the same concept [133]. This means the input queries of different users could be very different from each other. Second, some KGQA methods (e.g., [24, 165]) need to transform the natural language questions to graph queries, and then search the results according to these query graphs. The transformation algorithm may generate queries with inaccurate graph structure. Last but not the least, allowing users to input query graphs directly may introduce additional structural noise or inaccuracy due to their lack of full background knowledge of the underlying KG [133].

To address these issues, query ambiguity and vagueness need to be correctly resolved, which in turn requires new information in addition to the query itself. Relevance feedback (short for ReF) is one promising solution. The general idea behind relevance feedback is to take the results that are initially returned from a given query, to gather user feedback, and to use information about whether or not those results are relevant to form a new query.[1] The user feedback can be explicit (e.g., clicking like or dislike), implicit (e.g., the duration of time spent viewing a document) or pseudo feedback (no feedback is given). In spite of the fact that relevance feedback has been studied extensively and proven to be effective in information retrieval, it has not been well studied in graph query system. Due to the unique characteristics of graph query, traditional relevance feedback methods are not directly applicable. How to use relevance feedback to improve query performance on graph data largely remains an open problem.

In this chapter, we propose PREFNET which applies pseudo relevance feedback (short for PReF) to graph data to infer the true query structure from an ambiguous query and improve answering performance. More specifically, guided by Bayes rule, the proposed PREFNET decomposes the question answering problem into several components and models each component as a neural network. To infer the true query, PREFNET utilizes variational Bayesian inference. To boost the quality of the inferred query, a neighborhood embedding based VGAE model is used to prune inferior query. Finally, the newly inferred queries will be used to re-rank the original candidate answers. The experiment results show that our methods can achieve better question answering accuracy compared with the state-of-the-art baselines and better infer the user's query intention.

[1] https://en.wikipedia.org/wiki/Relevance_feedback.

7.2 Problem Definition

Table 7.1 gives the main notations used throughout this chapter. A knowledge graph can be denoted as $\mathcal{G} = (\mathcal{V}, \mathcal{R}, \mathcal{L})$ where $\mathcal{V} = \{v_1, v_2, ..., v_n\}$ is the set of nodes/entities, $\mathcal{R} = \{r_1, r_2, ..., r_m\}$ is the set of relations and $\mathcal{L}$ is the list of triples. Each triple in the knowledge graph can be denoted as (h, r, t) where $h \in \mathcal{V}$ is the head (i.e., subject) of the triple, $t \in \mathcal{V}$ is the tail (i.e., object) of the triple and $r \in \mathcal{R}$ is the edge (i.e., relation, predicate) of the triple which connects the head h to the tail t. For example, (`New_York`, `locatedIn`, `United_States`) and (`London`, `locatedIn`, `United_Kingdom`) are two triples in knowledge graph Yago.

Knowledge graph question answering aims to answer a question with the help of knowledge graphs. According to the study in [133], most users formulate queries using their own knowledge and vocabulary during the search process. They might not have a fairly good understanding of the underlying data schema and the knowledge graph structure. This means that the users' true intentions behind the queries may be frequently misinterpreted

Table 7.1 Notations and definitions

Symbols	Definition
$\mathcal{G} = (\mathcal{V}, \mathcal{R}, \mathcal{L})$	The knowledge graph
$\mathcal{V}$	The entity set
$\mathcal{R}$	The relation set
$\mathcal{L}$	The fact set
v_i	The ith entity/node in knowledge graph
r_i	The ith relation/edge in knowledge graph
$\mathbf{e}_{v_i}/\mathbf{e}_i$	The embedding of node v_i
$\mathbf{r}_i$	The embedding of relation r_i
Q	The ambiguous question
h_Q	The ambiguous question embedding
T	The true query graph of Q
$\overline{Q}$	The natural language question training set
A_Q	The answer set of question Q
a	A candidate answer of question Q
v_Q	The topic entity in question
z	The latent path
p_i	The ith relation in latent path
w_i	The ith word in Q
r_Q	The relation of query

or misrepresented. For example, if a user inputs query "thomas jefferson role in declaration independence", the phrase "role in" should be interpreted as predicate "profession" or "occupation", and the phrase "declaration independence" refers to the pronouncement and founding document adopted by the Second Continental Congress. While in query "John Litel role in declaration independence", the phrase "role in" should be interpreted as predicate "film actor" or "act in", and the phrase "declaration independence" represents a movie. The same phrases "role in" and "declaration independence" have totally different interpretations in the two queries. Unless the transformation algorithm can accurately understand the user's intention, directly answering these ambiguous queries may yield unwanted answers and deteriorate user experience. Here, a promising way to disambiguate the query is via relevance feedback as follows. With the help of relevance feedback, we can infer from the top rank candidates (e.g., `White_House`, `Politician`, `President` for "thomas jefferson" and `United_States`, `Actor`, `Albany_Wisconsin` for "John Litel") that "thomas jefferson" is a statesman and "John Litel" is a actor. These candidates provide clear and useful interpretation to users. The new interpretation can not only help users understand the background knowledge graph better, but also return more accurate answers. More examples can be found in Sect. 7.4.5B.

In this chapter, we focus on answering ambiguous one-hop question over knowledge graph. We assume the input ambiguous query Q contains a topic/anchor entity $v_Q \in \mathcal{V}$ and a sequence of words $Q = (w_1, w_2, ..., w_{|Q|})$. Ideally, each question can be mapped to a unique relation r_Q in the knowledge graph. The goal of question answering over knowledge graph is to identify a set of nodes $A_Q \subseteq \mathcal{V}$ which can answer the ambiguous question. We assume that all the answer entities exist in the knowledge graph, each question only contains a single topic/anchor entity $v_Q \in \mathcal{V}$ and v_Q is given.

To disambiguate the query and improve query accuracy, a natural idea is to effectively acquire more query-specific information. This can be achieved by relevance feedback which has been extensively studied in Information Retrieval Systems. In this chapter, we focus on utilizing pseudo relevance feedback to infer true query intention and improve question answering accuracy. More specifically, the proposed PREFNET treats the top-k candidate answers of a KGQA system as most relevant information and uses them to derive the true query so that more accurate answers can be found. The derived queries will in turn help the users better formulate their queries and obtain better understanding of the background knowledge.

Formally, the problem this chapter studies is defined as follows.

Problem 7.1 Answering Ambiguous Query:

Given: (1) A knowledge graph $\mathcal{G}$, (2) an ambiguous one-hop natural language question;

Output: (1) The answer of the question, (2) Top-k most likely correct query relations of the input query.

7.3 Proposed Method

In this section, we first introduce the framework and basic idea behind our method. Then, the details of each specific component are elaborated.

7.3.1 Model Overview

The key idea of PREFNET is as follows. We treat the top-k results of the KGQA system as potentially correct answers, and use them to predict the true query relation. Given the ambiguous query and its anchor entity, we give the following lemma to decompose the problem of question answering over knowledge graph (KGQA).

Lemma 7.1 (KGQA Decomposition) *Given an ambiguous query Q and its anchor node v_Q, let $Pr(T|Q, v_Q)$ denote the probability that query relation T is generated from Q and let $Pr(a|T, v_Q)$ denote the probability that candidate answer a found by T is the true answer, we have $Pr(T|Q, v_Q)Pr(a|T, v_Q) \propto Pr(Q|T, v_Q)Pr(T|a, v_Q)Pr(a|v_Q)$.*

Proof The probability of inferring the true query relation can be expressed as the following equation according to Bayes rule:

$$\begin{aligned} Pr(T|Q, v_Q) &= \frac{Pr(Q, T|v_Q)}{Pr(Q|v_Q)} \\ &\propto Pr(Q, T|v_Q) \\ &= Pr(Q|T, v_Q)Pr(T|v_Q), \end{aligned} \tag{7.1}$$

where $Pr(Q|v_Q)$ can be ignored since it is irrelevant to inferring T. Furthermore, given a candidate answer a of ambiguous query Q, we have

$$Pr(a|T, v_Q) = \frac{Pr(a, T|v_Q)}{Pr(T|v_Q)}, \tag{7.2}$$

This means

$$\begin{aligned} Pr(T|v_Q) &= \frac{Pr(a, T|v_Q)}{Pr(a|T, v_Q)} \\ &= Pr(a|v_Q)\frac{Pr(T|a, v_Q)}{Pr(a|T, v_Q)}. \end{aligned} \tag{7.3}$$

Plugging Eqs. (7.3)–(7.1), we can get

$$Pr(T|Q, v_Q) \propto Pr(Q|T, v_Q)Pr(a|v_Q)\frac{Pr(T|a, v_Q)}{Pr(a|T, v_Q)}. \tag{7.4}$$

Consequently, we obtain

$$Pr(T|Q, v_Q)Pr(a|T, v_Q) \propto Pr(Q|T, v_Q)Pr(T|a, v_Q)Pr(a|v_Q). \tag{7.5}$$

□

In Eq. (7.5), $Pr(a|v_Q)$ (**prior of candidate answer**) can be treated as the prior probability that node a is the correct answer given v_Q. It describes experts' beliefs when giving no evidence. The intuition is that if a is close to v_Q (e.g., the shortest distance of a to v_Q on the KG is small), it may have a high probability to be the answer. On the other hand, if a is far from v_Q, its probability to be the answer will be low. $Pr(Q|T, v_Q)$ (**likelihood of ambiguous query**) refers to the probability that given the anchor entity v_Q and the true query relation T, how likely the users will input ambiguous query Q. In general, if Q is similar to T, $Pr(Q|T, v_Q)$ should be high. Otherwise, it should be low. $Pr(T|a, v_Q)$ (**posterior of true query**) presents that given v_Q and candidate answer a, how likely T is the correct relation that reflects the true relationship between them.

The left hand side of Eq. (7.5) shows the main idea of question answering over knowledge graph. The KGQA system first transforms the input natural language question Q to a high quality query relation T, then finds the answer according to T and the anchor node v_Q. The goal of KGQA is to maximize $Pr(T|Q, v_Q)Pr(a|T, v_Q)$, which is equivalent to find query relation T which can maximize $Pr(Q|T, v_Q)Pr(a|v_Q)Pr(T|a, v_Q)$. However, directly calculating Eq. (7.5) is impossible since we do not know the real world distribution. A general idea is to use $Pr(T|a, v_Q)$ (**posterior of true query**) predict the "hidden relevance" between candidate a and anchor node v_Q. If the "hidden relevance" T is highly related to the original query Q, it can be used to re-rank the top candidate answers. Otherwise, the "hidden relevance" T is considered as noise and ignored.

More specifically, two neural network components are used in PrefNet. The first neural network is used to model $Pr(T|a, v_Q)$ (**posterior of true query**) which performs like a generator to produce high quality candidate query relations (the "hidden relevance" T). The second neural network is used to model $Pr(Q|T, v_Q)$ (**likelihood of ambiguous query**) which behaves like a discriminator to measure the similarity between Q and T, and prunes T if the similarity is too low.

Figure 7.1 shows the Framework of PrefNet. The yellow part corresponds to the method EmbedKGQA [126].[2] In the first step, EmbedKGQA is used to find k top candidates based on the input ambiguous query Q. Then, the query inference part ($Pr(T|a, v_Q)$ Sect. 7.3.2) is used to infer the true relation between anchor node v_Q and each candidate node e_i. Because not every candidate node is good, to boost the quality of those inferred relations, we use a neighborhood embedding based VGAE [66] model to prune low quality relations. Finally, the rest of the high quality inferred relations will be used by Query Ranking ($Pr(Q|T, v_Q)$ Sect. 7.3.4) and Answer Re-ranking (Sect. 7.3.5).

[2] We use EmbedKGQA because it's one of the state of the art methods.

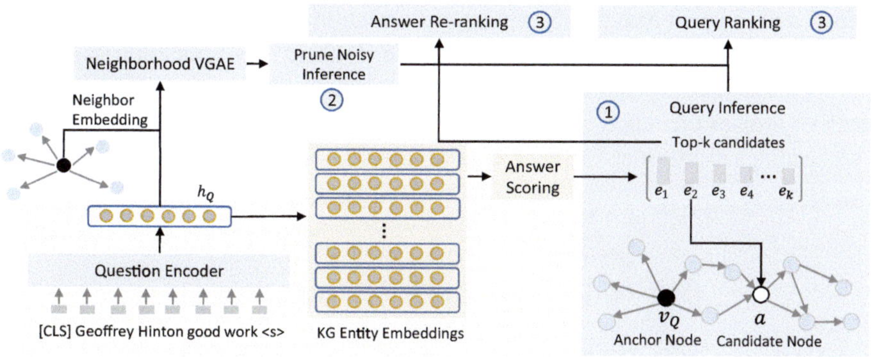

Fig. 7.1 The Framework of PREFNET. The yellow part shows the architecture of EmbedKGQA [126]

After we have introduced the general idea of PREFNET, in the following subsections, we introduce how to model each component in detail.

7.3.2 Query Inference: Posterior of True Query

When using pseudo relevance feedback to find out the true intention of the input query, we assume the top-k candidate answers returned by the existing algorithm in the first round is an initial set of most relevant answers. To better understand the query intention, PREFNET infers the query structure according to the anchor entity v_Q, potential/candidate answers, and the background knowledge graph.

When the input query is a one-hop query, this problem is equivalent to the link prediction problem in the knowledge graph. Given an entity pair (v_Q, a), we assume that there is a latent measure for a given entity pair in the KG, i.e. the collection of linked paths; these hidden information can reveal the underlying semantics between these two entities. More specifically, the posterior of the true query $\log Pr(T|a, v_Q)$ can be calculated by

$$\log Pr(T|a, v_Q) = \log \sum_{z} Pr_\theta(z|a, v_Q) Pr_\eta(T|z, a, v_Q) \tag{7.6}$$

where η and θ are model parameters, z is the latent path between v_Q and a in the knowledge graph. $Pr_\theta(z|a, v_Q)$ can be treated as the prior distribution denoting how likely z can represent the relation between a and v_Q, and $Pr_\eta(T|z, a, v_Q)$ can be treated as the likelihood that T is the true query between a and v_Q given latent variable z.

A—Estimate $Pr_\theta(z|a, v_Q)$. When calculating $Pr_\theta(z|a, v_Q)$, neighborhood of anchor node v_Q and candidate answer a usually provides us extra information which can imply their similarity to z. For example, if v_Q is surrounded by relations like `bornAt`, `liveIn`, `hasChild`, etc., it alludes that v_Q is a person. And if a is surrounded by `hasMajor`, `locatedIn`, `university_founder`, etc., we can infer that a might be a university.

Therefore, if z is a latent path which expresses the relation between a person and a university, $Pr_\theta(z|a, v_Q)$ should have a high value. Otherwise, $Pr_\theta(z|a, v_Q)$ should be low.

In order to encode the neighborhood information of anchor node v_Q and candidate answer a, PREFNET utilizes the relation/edge message passing graph neural network [150] to learn their embeddings as follows

$$m_v^k = \sum_{e \in N(v)} s_e^k, \tag{7.7}$$

$$s_e^{k+1} = \sigma([m_{v_Q}^k, m_a^k, s_e^i]W^k + b^k), v_Q, a \in N(e), \tag{7.8}$$

where $[\cdot]$ is the concatenation function, v is an arbitrary node in the knowledge graph, W^k and b^k are the learnable transformation matrix and bias, respectively.[3] s_e^k is the embedding of edge e at iteration k, s_e^0 is the initial edge embedding $\mathbf{r_e}$. Relation/Edge message passing is repeated for $\mathbf{K}$ times. The final message $m_{v_Q}^{\mathbf{K}-1}$ and $m_a^{\mathbf{K}-1}$ are taken as the representation for anchor node v_Q and candidate answer a, respectively.

On the other hand, given a path $z = (p_1, p_2, ..., p_{|z|})$ in the knowledge graph, the path embedding can be calculated by an LSTM as follows:

$$\mathbf{Z} = \text{LSTM}(z), \tag{7.9}$$

where $\mathbf{Z}$ denotes the embedding with all relational information aggregated throughout path z. The neighborhood information of a and v_Q can be aggregated by

$$\begin{aligned} s_{(v_Q,a)} &= \text{MLP}(m_{v_Q}^{\mathbf{K}-1}, m_a^{\mathbf{K}-1}) \\ &= \sigma([m_{v_Q}^{\mathbf{K}-1}, m_a^{\mathbf{K}-1}]W^{\mathbf{K}-1} + b^{\mathbf{K}-1}). \end{aligned} \tag{7.10}$$

Finally, the probability $Pr_\theta(z|a, v_Q)$ can be measured by the similarity between $\mathbf{Z}$ and $s_{(v_Q,a)}$:

$$Pr_\theta(z|a, v_Q) = \text{Sigmoid}(s_{(v_Q,a)} \cdot \mathbf{Z}), \tag{7.11}$$

where $\cdot$ is the dot product. An illustrative example is given in Fig. 7.2.

B—Estimate $Pr_\eta(T|z, a, v_Q)$. For likelihood $Pr_\eta(T|z, a, v_Q)$, it can be modeled by a graph decoder, which takes the embeddings of z, v_Q and a as its initial input and generates a relation. It can be simply modeled as:

$$Pr_\eta(T = r_j|z, a, v_Q) = \text{softmax}(\text{FNN}(s_{(v_Q,a)}||\mathbf{Z}))[j], \tag{7.12}$$

where FNN is the feedforward neural network.

[3] Graph neural network [174, 175] is a commonly used way to learn node representations.

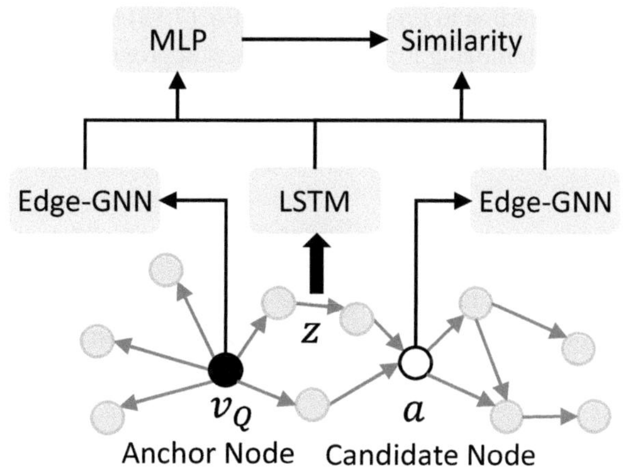

Fig. 7.2 Using relation/edge message passing to estimate $Pr_\theta(z|a, v_Q)$

7.3.3 Query Inference Training

Directly optimizing the objective in Eq. (7.6) is challenging, because the summation over all possible paths is intractable. In order to maximize Eq. (7.6), we resort to variational inference and minimize its negative evidence lower bound:

Proposition 7.1 *Let* $\mathcal{L}(\phi, \theta, \eta)$ *be defined as in Eq.* (7.13)*, we have* $\log Pr(T|a, v_Q) \geq -\mathcal{L}(\phi, \theta, \eta)$

$$\begin{aligned}\mathcal{L}(\phi, \theta, \eta) = & E_{z \sim q_\phi(z|T, a, v_Q)}[-\log Pr_\eta(T|z, a, v_Q)] \\ & + KL(q_\phi(z|T, a, v_Q)|Pr_\theta(z|a, v_Q))\end{aligned} \tag{7.13}$$

where $q_\phi(z|T, a, v_Q)$ *is the variational posterior.*

Proof According to Eq. (7.6), we have

$$\begin{aligned}\log Pr(T|a, v_Q) &= \log \sum_z Pr_\theta(z|a, v_Q) Pr_\eta(T|z, a, v_Q) \\ &= \log \sum_z q_\phi(z|T, a, v_Q) \frac{Pr_\theta(z|a, v_Q) Pr_\eta(T|z, a, v_Q)}{q_\phi(z|T, a, v_Q)} \\ &\geq \sum_z q_\phi(z|T, a, v_Q) \log \frac{Pr_\theta(z|a, v_Q) Pr_\eta(T|z, a, v_Q)}{q_\phi(z|T, a, v_Q)} \\ &= E_{z \sim q_\phi(z|T, a, v_Q)}[\log Pr_\theta(z|a, v_Q) \\ &+ \log Pr_\eta(T|z, a, v_Q) - \log q_\phi(z|T, a, v_Q)] \\ &= -\mathcal{L}(\phi, \theta, \eta)\end{aligned} \tag{7.14}$$

□

According to Proposition 7.1, Eq. (7.13) is the negative lower bound of Eq. (7.6). Therefore, minimizing Eq. (7.13) can best approximate $\log Pr(T|a, v_Q)$.

When approximating the posterior $q_\phi(z|T, a, v_Q)$, we use the same model as prior (Eq. (7.11)), except that we replace Eq. (7.11) by

$$\begin{aligned} x &= s_{(v_Q,a)} + e_T \\ q_\phi(z|T, a, v_Q) &= \text{Sigmoid}(x \cdot \mathbf{Z}) \end{aligned} \tag{7.15}$$

where e_T is the embedding of T, and $\cdot$ is the dot product.

When updating the model, we leverage gradient descent to minimize the loss. The gradient of the loss w.r.t. $q_\phi(z|T, a, v_Q)$ can be calculated as:

$$\begin{aligned} \nabla_\phi \mathcal{L} = E_{q_\phi(z|T,a,v_Q)}[\nabla_\phi q_\phi(z|T, a, v_Q)(\log Pr_\theta(z|a, v_Q) \\ + \log Pr_\eta(T|z, a, v_Q) - \log q_\phi(z|T, a, v_Q))]. \end{aligned} \tag{7.16}$$

The gradient in Eq. (7.16) can be approximated by the Monte Carlo method using N samples of the latent variable from variational distribution

$$\begin{aligned} \nabla_\phi \mathcal{L} \approx \frac{1}{N} \sum_{i=1}^{N} [\nabla_\phi q_\phi(z_i|T, a, v_Q)(\log Pr_\theta(z_i|a, v_Q) \\ + \log Pr_\eta(T|z_i, a, v_Q) - \log q_\phi(z_i|T, a, v_Q))]. \end{aligned} \tag{7.17}$$

When updating other parameters: θ, η, the gradient of $\mathcal{L}(\phi, \theta, \eta)$ with respect to η can be calculated directly, and the gradient of $\mathcal{L}(\phi, \theta, \eta)$ with respect to θ is equivalent to calculating the gradient of KL-divergence of $KL(q_\phi(z|T, a, v_Q)|Pr_\theta(z|a, v_Q))$ with respect to θ.

7.3.4 Query Ranking: Likelihood of Ambiguous Query

After we introduce how to model $Pr(T|a, v_Q)$ (**posterior of true query**), we now introduce how to model $Pr(Q|T, v_Q)$ (**likelihood of ambiguous query**).

Given the initial top-k answer candidates, a set of potential true query relations are generated by $Pr(T|a, v_Q)$ (**posterior of true query**). Because noise may exist in the initial candidate set, to boost the qualify of the new generated T, we use a neighborhood embedding based VGAE [66] to prune low quality T.

After pruning low qualify inferred queries, we use another neural network to model $Pr(Q|T, v_Q)$ which calculates the similarity between T and the input ambiguous query Q. Inspired by [11], $Pr(Q|T, v_Q)$ is approximated according to their distance in the embedding space. Given a ambiguous query, since the query is a natural language question, we use a pre-trained BERT [37] to learn its embedding:

$$[\mathbf{h}_{CLS}, \mathbf{w}_1, ..., \mathbf{w}_{|Q|}, \mathbf{h_s}] = \text{BERT}([CLS], w_1, ..., w_{|Q|}, < s >), \tag{7.18}$$

where $\mathbf{h}_{CLS}$ is the embedding of the [CLS] token and $\mathbf{h}_\mathbf{s}$ is the embedding of the $< s >$ token. The final question embedding is obtained from $\mathbf{h}_{CLS}$ as below, where FNN is a feed forward neural network

$$\mathbf{h}_Q = \text{FNN}([\mathbf{h}_{CLS}]). \tag{7.19}$$

When training the Query Ranking model $(Pr(Q|T, v_Q))$, we randomly sample some negative relations from the knowledge graph. The positive query relations are given in the training data. In summary, given a positive pair and a negative pair, the loss is defined as

$$L = \sum_{e_T} \sum_{e_{T'}} [\gamma + d(h_Q, e_T) - d(h_Q, e_{T'})], \tag{7.20}$$

where $e_{T'}$ is the embedding of the negative sample T', $d(h_Q, e_T)$ measures the distance between embedding h_Q and e_T, which can be either the L_1 or the L_2-norm, and γ is a margin hyperparameter.

During the test, we use the neighborhood embedding based VGAE to measure the quality of the inferred queries. If the score is greater than 0.5, we think the queries have high quality and return them directly. Otherwise, all the high qualify potential queries generated by query inference $Pr(T|a, v_Q)$ are ranked according to their distances calculated by $d()$ in Eq. (7.20).

7.3.5 Answering Re-ranking

After getting high quality inferred queries, we can re-rank all the top-*k* candidate answers. The ranking function is calculate by $\sum_T Pr(a|T, v_Q)$. Similar to [126], we approximate $\sum_T Pr(a|T, v_Q)$ as

$$\log \sum_T Pr(a|T, v_Q) \propto \gamma * |R_a \cap R_T|, \tag{7.21}$$

where R_a is the set of relations on the shortest path between anchor entity v_Q and candidate answer a. R_T is the top-ranked potential queries generated by query inference $Pr(T|a, v_Q)$ (**posterior of true query**). To keep the high quality of the inferred query T, we prune those queries which are scored less than 0.2 by the neighborhood embedding based VGAE [66] model.

7.4 Experiments

In this section, we evaluate the performance of the proposed PREFNET on several public datasets. We first introduce the datasets and baselines used in the chapter, and then present the experiment results.

7.4.1 Experimental Setting

Three datasets are used in the chapter which are summarized below:

- **WebQuestionsSP** contains both the 1-hop questions and 2-hop natural language questions. There are 4,000 questions in total and all questions can be answered by Freebase.
- **SimpleQuestions** contains more than 100 K simple 1-hop natural language questions which can be answered by Freebase.
- **MetaQA** is a dataset on movie domain which contains more than 400 K natural language questions. The background knowledge graph contains information about directors, movies, genres and actors. We use the 1-hop questions in our experiment.

In the experiment, we test the effectiveness of PREFNET on complete KG and incomplete KG with 50 and 20% missing edges respectively. All missing edges are randomly deleted.

We test the query ranking performance on 4 baselines, including

- HGNet [24] uses a generation method to transform the input query to a SPARQL query and uses it to find the results.
- RnG-KBQA [165] adopts a bootstrapping strategy to train a query graph ranker and generator, and searches the background knowledge graph according to the generated query.
- VGAE is a variant of variational autoencoder [66]. We model the encoder component with a pre-trained Bert [37] and decoder component with a feedforward neural network. No neighborhood information is used here.
- Neighborhood VGAE is the relation pruning model used in the proposed PREFNET which has been introduced in Sect. 7.3.4. The details are given in Appendix.

We test question answering performance on 3 baselines, including:

- GraftNet [136] finds a question-specific subgraph containing KG facts, and then uses a graph neural network to predict the answers.
- Key-Value Memory Network (KVMem) [103] maintains a memory table which stores KG facts and uses this for retrieval.

- EmbedKGQA [126] conducts multi-hop reasoning through matching pre-trained entity embeddings with question embedding obtained from RoBERTa [37]. We use Embed-KGQA with relation matching which has better performance compared with the Embed-KGQA without relation matching.

Another related method is GRF [133]. However, neither the code nor the datasets are publicly available, and thus we leave the comparison with GRF to future work.

7.4.2 Performance of Query Ranking

Traditional KBQA methods usually transform the natural language query to a query graph, and then find the answer according to the query graph. However, because of the ambiguity in the input query, the generated query graph is usually inaccurate. The pseudo relevance feedback, on the other hand, can infer queries according to the top candidate answers. We run the experiments on datasets WebQSP and SimpleQA. Because the data format of the background knowledge graph of MetaQA is different from that required by HGNet and RnG-KBQA, we omit the query ranking experiment on MetaQA.

Following the same setup as in [126], we evaluate the accuracy using the Hit@1 metrics which is the fraction of times a correct answer was retrieved within the top-1 positions. Table 7.2 shows the results of Query Ranking. Four different methods have been used in the experiments. HGNet and RnG-KBQA are knowledge graph question answering algorithms. They first transform the ambiguous natural language question to a query graph, and then search the knowledge graph according to the generated query graph. VGAE is a variational autoencoder and decoder method. Neighborhood-VGAE is the proposed pruning method used in PREFNET.

As observed in Table 7.2, the proposed PREFNET has the highest query ranking accuracy compared with other baselines. For Webqsp, the performance boosts are 5.0 and 5.9% with 50 and 80% of the underlying knowledge graph respectively, compared with VGAE, and 0.9% on average compared with Neighborhood-VGAE. Similar results can be found in complete knowledge graph as well. Besides, compared with 50% knowledge graph, the

Table 7.2 Query ranking Hit@1

	Webqsp half	Webqsp 80% KG	Webqsp full	SimpleQA half	SimpleQA 80% KG	SimpleQA full
RnG-KBQA	61.2	63.3	66.7	–		
VGAE	70.7	71.3	73.7	87.4	89.1	91.6
Neighborhood-VGAE	74.2	76.1	78.1	90.3	92.3	93.5
PREFNET	75.7	77.2	79.0	90.8	92.7	94.5

query ranking accuracy on 80% and full knowledge graph of all four baselines is higher. This means if the background knowledge graph becomes more complete, all baselines tend to generate more accurate query graphs.

7.4.3 Performance of Question Answering

Table 7.3 shows the results of all baseline methods on Webqsp dataset with 50% complete KG, 80% complete KG and complete KG, respectively. As shown in Table 7.3, the Hit@1 of all methods decreases as the sparsity (incompleteness) of the background knowledge graph increases. This means that the quality of the background knowledge graph has a significant impact on the KGQA task. Among all the methods, EmbedKGQA with relation matching can achieve the highest accuracy. PREFNET further increases the accuracy by 1% on average. Tables 7.4 and 7.5 show the performance of different methods on SimpleQuestions and MetaQA datasets. Similar results can be observed.

7.4.4 Efficiency

Figure 7.3 shows the training time and test time of the query inference component on different datasets. As we can see, the runtime of the Query Inference component on Webqsp and SimpleQA is much larger than that on MetaQA dataset. This is because the background knowledge graph of Webqsp and SimpleQA is much larger than that of MetaQA. Despite the long training time, the test time of the Query Inference component is relatively short.

Compared with the running time of query inference, the running time of the query ranking (Sect. 7.3.4) and answer re-ranking (Sect. 7.3.5) is negligible (e.g., less than 20 min on the whole test dataset).

Table 7.3 KGQA Hit@1 results of Webqsp on complete, 50 and 80% completeness knowledge graphs

Model	Webqsp 50%	Webqsp 80%	Webqsp
GraftNet	32.7	39.8	–
KVMem	44.2	–	46.7
HGNet	37.8	45.4	46.1
RnG-KBQA	40.6	48.5	53.1
EmbedKGQA	46.3	52.8	56.7
PREFNET	*47.7*	*53.7*	*57.2*

Table 7.4 KGQA Hit@1 results of SimpleQA on complete, 50 and 80% completeness knowledge graphs

Model	simpleQA50%	simpleQA80%	simpleQA
GraftNet	39.8	–	–
KVMem	28.9	–	–
EmbedKGQA	46.5	58.9	64.3
PREFNET	*48.8*	*60.1*	*64.8*

Table 7.5 KGQA Hit@1 results of MetaQA on complete, 50 and 80% completeness knowledge graphs

Model	MetaQA 50%	MetaQA 80%	MetaQA
GraftNet	64.0	–	–
KVMem	63.6	–	–
EmbedKGQA	83.1	90.1	95.3
PREFNET	*83.9*	*90.6*	*95.6*

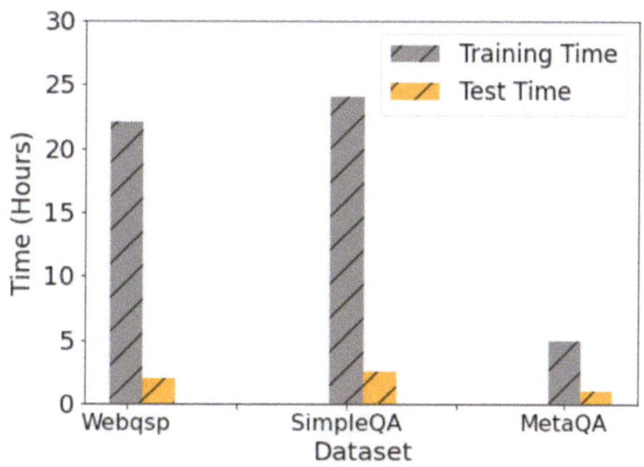

Fig. 7.3 Query inference training and test time

7.4.5 Ablation Study

A—Query Inference. In this subsection, we show the effectiveness of the query inference module. We first pretrain the module only on the background knowledge graph of each dataset, and then retrain the module on the question training dataset. During the pretrain and retrain processes, we split all the data into training set (80%), valid set (10%) and test set

Table 7.6 Query Inference Performance.wsp means WebQuestionsSP.sQA means SimpleQuestions

Model	wsp50%	wsp80%	sQA50%	sQA80%	MetaQA50%	MetaQA80%
Hit@1	82.4	87.4	92.6	95.9	86.2	88.4
Hit@3	91.8	93.4	98.8	99.6	98.1	93.2
Hit@5	93.7	94.7	99.3	99.9	99.9	94.1

(10%). As shown in Table 7.6,[4] the query inference module has a very high Hit@K on all datasets. It can achieve 88.8% Hit@1, 95.8% Hit@3 and 96.9% Hit@5 on average.

B—Query Ranking.

In this subsection, we show the effectiveness of the query ranking module. Some examples are shown in Table 7.7. As we can see, when the input question is ambiguous, it is very hard to correctly predict its true query intention. For example, for query "thomas jefferson role in declaration independence", the query generator VGAE thinks the query intention is "film.actor.film". Because the phrase "role in" is usually used to describle "someone plays/acts a role in a movie". However, with pseudo relevance feedback, the query inference module can discover "thomas jefferson" is a stateman/diplomat, and that helps the module correctly predict the true intention "people.person.profession" and finds the correct answer (statesperson) in the background knowledge graph.

C—Question Answering.

Table 7.8 shows the ablation study on question answering task. EmbedKGQA—RM stands for EmbedKGQA without relation matching. EmbedKGQA+QI denotes that we add the query information inferred by query inference module to relation matching process in EmbedKGQA. More specifically, the relation matching process of EmbedKGQA finds the shortest path between the anchor node v_Q and candidate answer a, and order the candidate answers according to their shortest paths. We add the relation inferred by query inference module to the shortest path when ordering the candidate answer. It is observed that the proposed PREFNET achieves the best performance.

[4] For the full background knowledge graph, because we can directly find correct paths between anchor node v_Q and candidate answer node a, we do not need query inference module.

Table 7.7 Results of relation found by different methods

Question	Query relation found by PrefNet	Query relation found by VGAE
Thomas jefferson role in declaration independence	people.person.profession	film.actor.film
State mount st. helens in	location.location.containedby	geography.mountain.mountain_type
Monarchy japan	location.country.form_of_government	schema.administrative_area.administrative_children
First gulf war fought	time.event.locations	military_command.military_commander
Buddha come	people.person.place_of_birth	people.person.places_lived
Part country new england	location.location.containedby	base.locations.countries.continent
Drink john pemberton create	inventor.inventions	symbols.name_source.namesakes
Jesus after he died cross	people.deceased_person.place_of_death	people.person.profession
Charles babbage discover	inventor.inventions	base.argumentmaps.innovator.original_ideas
Country code mexico	location.country.internet_tld	location.location.adjoin_s

Table 7.8 Ablation study of question answering

Model	webqsp50%	webqsp80%	simpleQA50%	simpleQA80%
EmbedKGQA—RM	44.8	50.3	46.5	58.9
EmbedKGQA	45.9	52.8	46.6	58.8
EmbedKGQA+QI	46.3	52.8	46.7	58.9
PREFNET	*47.7*	*53.7*	*48.8*	*60.1*

7.5 Discussion

Reranking candidate answers using symbolic subgraph information in the knowledge graph has shown promising results in this study. Integrating graph neural network-based models can better predict the true query intention, demonstrating the effectiveness of neural-symbolic reasoning. However, the current approach only utilizes explicit triplet information from the knowledge graph, missing other implicit semantic information from the question. Leveraging this implicit information is a promising future direction.

8 Ambiguous Entity Matching with Neural Symbolic Reasoning Over Incomplete KG

Entity Matching (EM) is a significant task involving the determination of the logical relationship between two entities, such as `Same`, `Different`, and `Ambiguous`. Traditional approaches to entity matching (EM) heavily depend on supervised learning, which necessitates a vast collection of high-quality labeled data. This labeling process is both time-consuming and costly, limiting the practical application of these methods. Consequently, there is a pressing demand for low-resource EM solutions that can perform effectively with minimal labeled data. Recently, prompt tuning-based approaches have shown promising results on low-resource entity matching, but they tend to focus solely on entity-level matching, overlooking crucial attribute-level information. Moreover, they lack interpretability and explainability. To address this limitation, in this chapter, we introduce PromptAttrib, a comprehensive solution that tackles entity matching challenges through attribute level prompt tuning and logical reasoning. PromptAttrib leverages both entity-level and attribute-level prompts to enhance matching accuracy by incorporating valuable contextual information, and it induces the matching label by fuzzy logic formulas. By considering attributes, the model gains a deeper understanding of the entities, leading to improved matching results. Moreover, PromptAttrib incorporates dropout-based contrastive learning on soft prompts, inspired by the SimCSE technique. This enhances the robustness of the model by introducing dropout mechanisms during training, making it more resilient to variations in input prompts.

L. Liu and H. Tong, *Neural Symbolic Knowledge Graph Reasoning*, Synthesis Lectures on Computer Science, https://doi.org/10.1007/978-3-032-15858-1_8

8.1 Introduction

Entity Matching (EM) is a fundamental problem in data management that focuses on determining whether two entity records refer to the same real-world entity. This task carries significant importance in various domains such as question answering [73], dialog system [1, 3, 146] and so on. In practical terms, consider a situation where two distinct knowledge databases store customer information from different sources. Due to differences in schema and data formats, effectively combining this information becomes a challenging task. However, by leveraging entity matching techniques, it becomes possible to identify matching customer records across knowledge databases, facilitating accurate data merging and linking. Thus, the application of entity matching plays a crucial role in improving data quality and enabling seamless information consolidation across diverse sources.

Conventional entity matching (EM) techniques rely extensively on abundant high-quality labeled data, which is often scarce or unavailable. Therefore, there is a growing need for low-resource EM methods that can achieve robust performance with minimal labeled data. Prompt tuning is a promising way for training and fine-tuning language models to tackle low-resource entity matching. By carefully crafting and adjusting the initial prompt given to the model, researchers and developers can guide the model's behavior and steer it towards desired outcomes. Existing prompt tuning based methods for low-resource entity matching, such as promptEM [151], commonly serialize each entity into a sentence, and a prompt template is used to generate the input for the language model. Despite these approaches have shown good performance in many scenarios, such prompt tuning based models function like a black-box and lack interpretability. Moreover, they consider only the entity-level information and overlook the crucial attribute information associated with each entity.

However, attribute information plays a significant role in accurately determining entity matches because it contains more fine-grained details. It can help us explain why two entities are the `Same`, `Different`, or `Ambiguous`. For example, Michael Jordan with the occupation "basketball player" is different from Michael Jordan with the occupation "computer scientist". Exploring innovative ways to incorporate attribute-level information holds promise for enhancing the robustness and reliability of entity matching systems.

In addition, existing prompt tuning methods typically assume that the input soft prompt is well-trained during the training process, without explicitly making the representations of different inputs distinguishable from each other. However, high-quality entity/attribute representations are crucial for the success of entity matching. If two entities/attributes are different, their representations should be easily distinguishable. Taking inspiration from SimCSE, a technique that utilizes dropout-based contrastive learning on language models to enhance representation learning quality, we propose applying dropout-based contrastive learning on soft prompts. This approach aims to increase the model's resilience to variations in input prompts by introducing dropout mechanisms during the training phase. By incorporating dropout-based contrastive learning on soft prompts, we seek to improve the overall performance of the entity matching model, ultimately leading to more accurate and consistent results.

In this chapter, we present PROMPTATTRIB, a comprehensive solution consisting of two main components. First, PROMPTATTRIB leverages both entity-level prompts and attribute-level prompts to effectively address the challenges related to low-resource entity matching. By incorporating fuzzy logic reasoning, the model predicts entity-level matches based on the logical relationships between different attribute information, making the prediction results more explainable. Second, PROMPTATTRIB employs dropout-based contrastive learning on soft prompts. This technique enhances the model's performance by encouraging distinguishable feature representations and reducing overfitting. Through dropout-based contrastive learning, the model learns to generalize better and captures more nuanced relationships between entities and attributes.

8.2 Problem Definition

In this section, we first present the problem definition of low-resource entity matching (EM). Next, we introduce the serializing method for EM, followed by an introduction of prompt-based tuning with LMs.

8.2.1 Problem Statement

Low-resource entity matching (EM) involves the identification of pairs of data entries from two collections that refer to the same real-world entity with limited labeled training datapoints. Formally, given two data sources E_A and E_B, we assign a binary label $y \in \{0, 1\}$ to each candidate pair $(e_a, e_b) \in E_A \times E_B$. Here, $y = 1$ indicates a true match, while $y = 0$ represents a mismatched pair.

Problem 8.1 Low-resource Entity Matching:

Given: (1) an entity e_a from data source E_A, (2) an entity e_b from data source E_B, (3) limited labeled training datapoints.

Output: a binary label $y \in \{0, 1\}$ for the entity pair (e_a, e_b)

8.2.2 Preliminary

Serializing: The matching problem can be effectively solved by formulating it as a sequence classification task. First, entity pairs are serialized to sequences, and then, a pre-trained LM is finetuned to solve the task. An entity with n attributes can be denoted

name	position	address	postalCode
McKees Rocks Bridge	40.47708; -80.04827	Mckees Rocks Bridge; Route 65	15233

Fig. 8.1 Example

as $e = \{\text{attr}_i, \text{val}_i\}_{i \in [1,n]}$, where attr_i is the attribute name and val_i is the corresponding attribute value. Then the serialization is denoted as:

$$\text{serialize}(e) ::= [\text{COL}]\text{attr}_1[\text{VAL}]\text{val}_1 \ldots [\text{COL}]\text{attr}_n[\text{VAL}]\text{val}_n \quad (8.1)$$

where [COL] and [VAL] are two special tags indicating the start of attribute names and values, respectively. Taking the relational entity in Fig. 8.1 as an example, we serialize it as:

$$[\text{COL}]\text{name}[\text{VAL}]\text{McKees Rocks Bridge} \ldots [\text{COL}]\text{postalCode}[\text{VAL}]15233 \quad (8.2)$$

8.2.3 Prompt-Based Tuning

Prompt-based tuning has been proposed to apply cloze-style tasks to tune LMs. Formally, we define a label word set $V_{\mathbf{y}} = \{w_1, \ldots, w_m\}$. $V_{\mathbf{y}}$ is a subset of the vocabulary V of the LM, i.e., $V_{\mathbf{y}} \subseteq V$. We get an overall dictionary V^* by taking the union of the dictionary corresponding to each label. Another primary component of prompt-tuning is a prompt template $T(\cdot)$, which modifies the original input x into a prompt input $T(x)$ by adding a set of additional tokens in x. Generally, a token $[MASK]$ is added for LMs to predict the missing label word $w \in V_{\mathbf{y}}$ using $T(x)$. Thus, in prompt-tuning, a classification problem is transferred into a masked language modeling problem: $p(y \in Y|x) = p([MASK] = w \in V_{\mathbf{y}}|T(x))$

In this section, we detail how to utilize prompt-tuning to deal with EM. We first design EM-specific prompt templates and label words, and then, we describe the training and inference process

8.2.4 Prompt Template

To cast the EM problem as a prompt-tuning one, we first design suitable prompt templates (i.e., hard-encoding templates and continuous templates) and label words set (to consider general binary relationship)

Hard-encoding Templates. For the choice of hard-encoding templates, we do not use automatic searching methods for discrete prompts since the EM task is clearly defined and the prompts are easily purposeful. Given each candidate pair $x = (e, e')$, we construct the following templates:

$$T_1(x) = \text{Are serialize}(e) \text{ and serialize}(e') \text{ the } [MASK] \quad (8.3)$$

$$T_2(x) = \text{serialize}(e) \text{ is } [MASK] \text{ to serialize}(e') \quad (8.4)$$

Note that various prompts can be used in the experiments, we provide only two simple examples here.

Continuous Templates. As prompt construction is to find a method that allows a LM to effectively perform a task, rather than being for human consumption, it is not necessary to limit the prompt to human-interpretable natural language. Thus, continuous prompts are proposed to perform prompting directly in the embedding space of the model. Here, we employ P-tuning, where continuous prompt tokens are learned by inserting trainable variables into the embedding input. Specifically, trainable prompt tokens are initialized, and then, BiLSTM is utilized to account for interaction between prompt tokens. This enables the model to find better continuous prompts beyond the original vocabulary V of model could express. We give an illustrative example in Fig. 8.2.

Label Words Set. In addition to designing templates, another primary component is to design the set of label words. Given two entities e_1 and e_2, it is often more informative to classify the relationship between them rather than simply determining if they are identical. Specifically, we aim to determine if the entities are relevant to each other, which encompasses a broader range of relationships beyond mere matching. For a general binary relationship, we map the label $y = \text{yes}$ into a set $V_y = \{\text{matched, similar, relevant}\}$. Similarly, the label $y = \text{no}$ is mapped to a set $V_y = \{\text{mismatched, different, irrelevant}\}$. This approach allows for a more nuanced classification that captures various degrees of relevance and irrelevance between the entities.

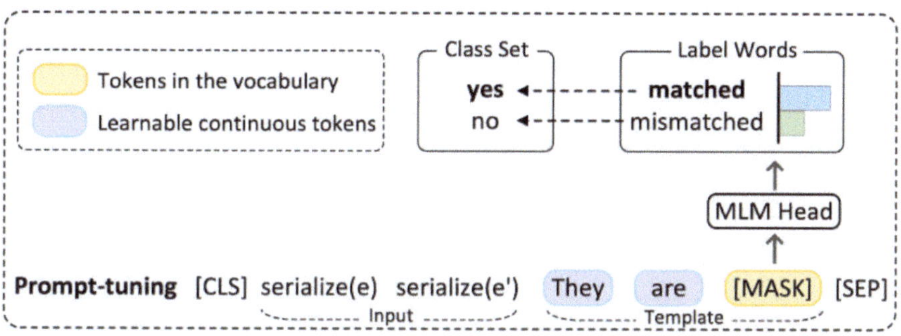

Fig. 8.2 The illustration of prompt-tuning. The blue rectangles in the figure are special prompt tokens, whose parameters are initialized and learnable during prompt-tuning. Image is from [151]

8.3 Method Details

8.3.1 Entity Level Prompt-Tuning

Inspired by existing prompt tuning-based methods, such as PromptEM [75, 151], we consider checking whether two entities are the same by prompt-tuning. More specifically, given two entities, we first serialize each entity into text using the method introduced in Sect. 8.2.4. Subsequently, these two serialized texts are used as input for the prompt template, as depicted in Fig. 8.3a. The resulting context is then fed as input to the large language model. The output embedding of the special token **[MASK]** is utilized by the **MLM Head** to predict the results. The label words set used in entity level prompt tuning is the same as the label words set introduced in Sect. 8.2.4. This means we map the label $y = \text{yes}$ to a set $V_y = \{\text{matched, similar, relevant}\}$. Similarly, the label $y = \text{no}$ is mapped to a set $V_y = \{\text{mismatched, different, irrelevant}\}$.

8.3.2 Attribute Level Prompt-Tuning

Existing prompt tuning-based methods primarily focus on considering only the entity-level information of two entities. However, attribute information is often of significant importance, as it can provide essential context to the system. To leverage this attribute information effectively, one potential approach is to integrate it into the prompt. In contrast to entity-level prompts that include only one **[MASK]** token, the attribute prompt contains both entity-level and attribute-level information, featuring multiple **[MASK]** tokens in the template. The first **[MASK]** is utilized to predict the entity-level information, while the subsequent **[MASK]** tokens are employed to predict the attribute-level information.

In most cases, this enhanced prompt with attribute information works effectively. However, certain entities may contain an extensive amount of information. For instance, the **Description** field of a product, such as a "computer," can be very lengthy, even exceeding 512 words. This situation poses a challenge as the combined length of the enhanced prompt with attribute information would exceed the model's maximum input

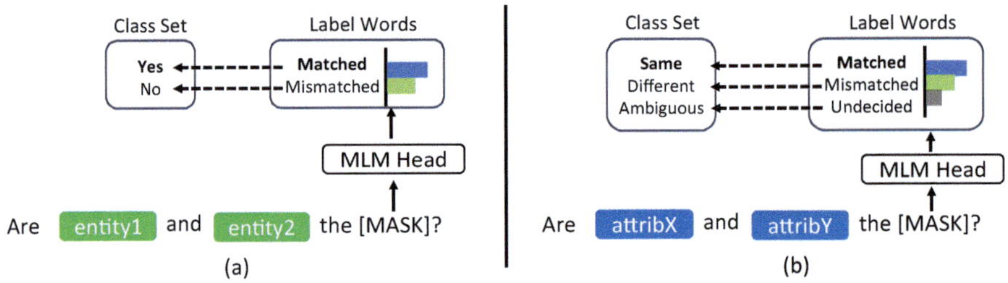

Fig. 8.3 The entity level prompt tuning and the attribute level prompt tuning

length, necessitating truncation of the text. To address this issue, we separate the entity-level prompt from the attribute-level prompt, as illustrated in Fig. 8.3b. Unlike entity-level prompt tuning, the label words set used in attribute-level prompt tuning contains three categories: `Same`, `Different`, and `Ambiguous`. More specifically, we map the label y = Same to a set V_y = {same, similar, positive}, the label y = Different is mapped to a set V_y = {mismatched, different, irrelevant}, and the label y = Ambiguous is mapped to a set V_y = {uncertain, unclear, neutral}.

8.3.3 Fuzzy Logic Reasoning

We observe the entity level matching can be logically induced from attribute level matching. Based on the definition of entity matching, we develop the following induction rules.
Same Rule: It is evident that if two entities are the same, then all their attributes should also be the same, implying that all paired attributes should be labeled as `Same`. Let $\{(a_k^1, a_k^2)\}_{k=1}^K$ be all attribute pairs. Then, we induce a attribute-level `Same` score by

$$S(\texttt{Same}|e_1, e_2) = \left[\prod_{k=1}^{K} P(\texttt{Same}|a_k^1, a_k^2)\right]^{\frac{1}{K}} \tag{8.5}$$

This works in a fuzzy logic fashion [52], deciding whether the entity-level label should be `Same` by considering the average of attribute level predictions. Here, we use the geometric mean because it is biased towards low scores. In other words, if there exists one attribute pair with a low `Same` score, then the chance of the entity label being `Same` is also low.
Difference Rule: Two entities are `Different` if there exists (at least) one paired attribute labeled as `Different`. The fuzzy logic version of this induction rule is given by

$$S(\texttt{Different}|e_1, e_2) = \max_{k=1,\ldots,K} P(\texttt{Different}|a_k^1, a_k^2) \tag{8.6}$$

Here, the max operator is used in the induction because the `Different` rule is an existential statement, i.e., there exist(s) $\cdots$.
Ambiguous Rule: Two entities are `Ambiguous` if there exists (at least) one `Ambiguous` attribute pair, but there does not exist any `Different` attribute pair. The fuzzy logic formula is

$$S(\texttt{Ambiguous} \mid e_1, e_2) = \left[\max_{k=1\ldots K'} P(\texttt{Ambiguous} \mid a_k^1, a_k^2)\right][1 - S(\texttt{Different} \mid e_1, e_2)] \tag{8.7}$$

The first factor determines whether there exists a `Ambiguous` attribute pair. The second factor evaluates the negation of "at least one different attribute" as suggested in the second clause of the Rule for `Ambiguous`.

Finally, we normalize the scores into probabilities by dividing the sum, since all the scores are already positive. This is given by

$$P(L|\cdot) = Z^{-1}S(L|\cdot) \tag{8.8}$$

where $L \in \{\texttt{Same}, \texttt{Different}, \texttt{Ambiguous}\}$, and Z is the normalizing factor which is defined as $Z = S(\texttt{Same}|\cdot) + S(\texttt{Different}|\cdot) + S(\texttt{Ambiguous}|\cdot)$.

During training process, we use cross-entropy loss to train our model by minimizing $-\log P(t|\cdot)$ where $t \in \{\texttt{Same}, \texttt{Different}, \texttt{Ambiguous}\}$ is the ground truth entity-level label. During the experiment, we categorize `Ambiguous` based on various datasets as either `Same` or `Different`.

8.3.4 Soft Token Contrastive Learning

Current methods for tuning prompts typically assume that the initial soft prompt is well-established during training and that the model's learned representation is sufficiently distinguishable. However, this assumption often does not hold, particularly in low-resource scenarios. High-quality representations of entities and attributes significantly influence performance. To tackle this issue, drawing inspiration from techniques such as SimCSE [48], we propose integrating contrastive learning into soft prompts to enhance representation learning.

The idea of contrastive learning is to enhance representation learning by training a model to distinguish between similar and dissimilar instances in the data. By maximizing the similarity of representations from similar instances and minimizing that of dissimilar ones, contrastive learning fosters the creation of more meaningful and discriminative feature representations.

In PromptAttrib, given an entity e_i, we construct its positive sampling data by randomly applying dropout masks to the soft prompt token inside serialize(e_i), resulting in e_i^+. This approach has been utilized in various models; for instance, in the standard training of Transformers [149], dropout masks are applied to fully-connected layers and attention probabilities (typically $p = 0.1$) to enhance model robustness. After obtaining the positive data point, we pass it through the prompt-tuning model $z = f_\theta(e_i^+)$ to obtain its embedding z. This process is repeated twice using different dropout masks to generate two positive embeddings z_1 and z_2, and the training objective becomes: $L = \|z_1 - z_2\|_2$

Note that unlike SimCSE, which applies dropout to the language model parameters, we keep the language model unchanged because modifying parameters in language models can be time-consuming. Instead, we apply dropout to the input soft tokens within the prompt. This approach ensures that even if the input prompt context remains constant, applying dropout

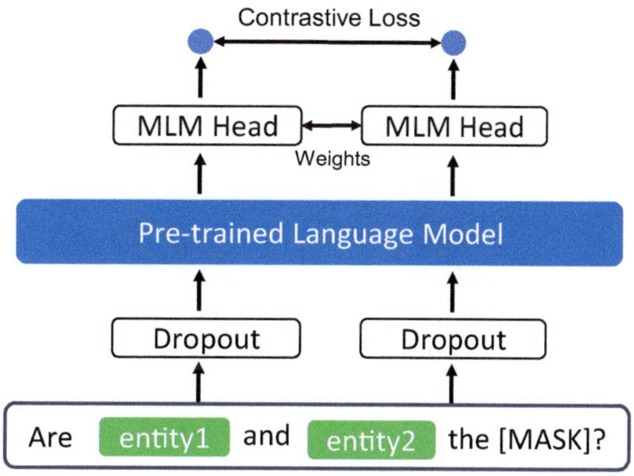

Fig. 8.4 Contrastive learning framework

multiple times yields distinct inputs. This concept bears similarity to siamese networks [116], albeit focusing solely on positive pairs in the contrastive loss. The final framework of the model is shown in Fig. 8.4.

8.4 Experiment

In this research, we conducted a comprehensive evaluation of our proposed method in the domain of entity matching and applied it to various datasets representing diverse real-world scenarios. The datasets used in our experiments are as follows:

- Geo-heter: The Geo-heter dataset encompasses a total of 194,089 geospatial entities sourced from well-known location-based platforms, including Yelp, Foursquare (FSQ), and OpenStreetMap (OSM).
- Cameras: The Cameras dataset is a subset of WDC dataset. The WDC dataset comprises an extensive collection of 26 million product offers and descriptions gathered from a wide range of e-commerce websites. The Cameras dataset focuses on entities related to cameras. It includes a variety of camera models, brands, specifications, and other camera-related attributes.
- Computers: The Computers dataset is also a subset of WDC dataset which comprises entities related to computers, incorporating various computer models, brands, configurations, and other computer-related information.
- ISWC (International Semantic Web Conference): The ISWC dataset consists of entities related to the International Semantic Web Conference, including papers, authors, venues, and publication years.

During the experiments, we use only 5% of the labeled training data to simulate the low-resource entity matching scenario. The following methods are employed as baselines.

- SentenceBERT [116]: SentenceBERT proposes a siamese architecture for pretrained LMs for sentence matching, and could also be applied to the task of GEM.
- DeepMatcher [156]: DeepMatcher is an entity matching framework that uses RNN architecture to aggregate the attribute values and then align the aggregated representations of the attributes.
- Ditto [75]: Ditto is one of the SOTA EM approach that fine-tunes a pre-trained LM with three optimizations (i.e., domain knowledge, TF-IDF summarization, and data augmentation).
- PromptEM [151]: PromptEM: PromptEM is another method for entity matching tasks, particularly designed for prompt-based tuning of language models. It applies the concept of cloze-style tasks to tune large language models for entity matching.

All experiments are conducted on a machine with an Intel(R) Xeon(R) Gold 6240R CPU, 1510 GB memory, and an NVIDIA-SMI Tesla V100-SXM2 GPU.

8.4.1 Entity Matching Performance

In this subsection, we test the performance of different baseline methods on entity matching task. We use metrics: F1-score, Average Precision and Accuracy to measure the performances of all the baselines.

Table 8.1 shows the results. As we can see, for the geo-heter dataset, PROMPTATTRIB outperforms both Ditto and PromptEM, achieving the highest F1-score (0.811), Average Precision (0.853), and Accuracy (0.889). PromptEM also performs well with an F1-score of 0.785, while Ditto has the lowest F1-score of 0.036.

In the cameras dataset, PROMPTATTRIB has the highest F1-score (0.455), Average Precision (0.456) and Accuracy (0.722), while PromptEM achieves the second highest Accuracy of 0.688 compared to other methods. For the computers dataset, PROMPTATTRIB again outperforms the other methods, achieving the highest F1-score (0.477), Average Precision (0.495) and Accuracy (0.721). PromptEM follows closely with an F1-score of 0.495 and an Average Precision of 0.452. Ditto lags behind with an F1-score of 0.449. In the ISWC dataset, PROMPTATTRIB also performs the best, obtaining the highest F1-score (0.775), Average Precision (0.796) and Accuracy (0.825). PromptEM closely follows with an F1-score of 0.764 and, Average Precision of 0.736 and Accuracy 0.819. Ditto achieves an F1-score of 0.721. Overall, the results indicate that PROMPTATTRIB consistently outperforms other baselines across the different datasets, showcasing its effectiveness in entity matching tasks.

Table 8.1 The entity matching performance of different methods. F means F1-score, P means average precision, A means Accuracy

Dataset	Geo-heter			Cameras			Computers			ISWC		
	F	P	A	F	P	A	F	P	A	F	P	A
SentenceBERT	58.8	53.8	64.8	38.3	21.4	21.3	39.4	35.6	57.5	66.2	68.0	75.5
DeepMatcher	43.8	28.9	72.1	39.2	22.4	23.0	42.8	37.1	58.9	69.1	70.6	76.3
Ditto	3.6	45.5	71.3	42.8	27.3	27.3	44.9	40.3	62.1	72.1	72.7	80.6
PromptEM	78.5	84.5	85.5	35.4	29.4	68.8	49.5	45.2	68.4	76.4	73.6	81.9
PROMPTATTRIB	*81.1*	*85.4*	*88.9*	*45.5*	*45.6*	*72.2*	*47.7*	*49.5*	*72.1*	*77.5*	*79.6*	*82.5*

Table 8.2 The entity matching performance of different language models as backbone

	F	P	A
Dataset	Geo-heter		
BERT	77.2	73.5	83.4
Roberta	79.5	78.2	86.4
Albert	78.7	80.2	86.6
GPT2	64.7	68.4	79.7
Roberta-large	81.1	85.4	88.9
Albert-large	79.9	83.1	86.2

8.4.2 Performance of Different Language Models

We also conducted experiments using various large language models as the backbone of the prompt tuning model. Specifically, we evaluated six models, including Bert [37], Roberta [94], Albert [72], GPT2 [115], Roberta-large, and Albert-large. The performance results are presented in Table 8.2.

As we can see, the results show that among the tested language models, Roberta-large achieved the highest F1-score (0.811), Average Precision (0.854) and Accuracy (0.889), closely followed by Albert-large (F1-score: 0.799, Average Precision: 0.831, and Accuracy: 0.862). Roberta also performs well with an F1-score of 0.795 and an Average Precision of 0.782. BERT obtained an F1-score of 0.772 and an Average Precision of 0.735. GPT2 achieved a lower F1-score of 0.647 and an Average Precision of 0.684 compared to the other language models. In terms of Accuracy, Roberta-large again stands out with a value of 0.889, followed closely by Albert (Accuracy: 0.866) and Albert-large (Accuracy: 0.862). Overall, the results demonstrate that language models, such as Roberta-large and Albert-large, tend to perform better than smaller ones like BERT and GPT2, indicating the importance of model size and capacity in improving entity matching performance.

8.4.3 Ablation Study

In this subsection, we examine the effectiveness of integrating contrastive learning with attribute-level prompt tuning. We evaluate how this combination improves model's overall performance, particularly in low-resource scenarios. By analyzing various metrics and comparing results under different scenarios, we aim to demonstrate their advantages.

Dropout ratio. Table 8.3 shows the entity matching performance with contrastive learning under different dropout ratios, where "PROMPTATTRIB 0.35" means the dropout ratio is 0.35 and "PROMPTATTRIB No" means without dropout. As we can see, PROMPTATTRIB with a dropout ratio of 0.35 achieves the highest F-score (0.455) and average precision (0.456),

Table 8.3 The entity matching performance with contrastive learning under different dropout ratios

	F	P	A
Dataset	Cameras		
Model	F	P	A
Ditto	42.9	27.3	27.3
PromptEM	42.3	31.7	71.3
PromptAttrib No	40.4	39.7	71.7
PromptAttrib 0.35	45.5	45.6	72.3
PromptAttrib 0.4	37.6	39.0	70.9
PromptAttrib 0.45	39.1	39.5	73.1

indicating that this ratio provides the best balance between regularization and information retention. Comparing "PromptAttrib No" and PromptAttrib with different dropout ratios, it is evident that applying dropout generally improves performance. For instance, "PromptAttrib No" achieves an F-score of 0.404, while "PromptAttrib 0.35" achieves a significantly higher F-score of 0.455. This improvement suggests that dropout helps in preventing overfitting and enhances the model's ability to generalize from the training data. Interestingly, PromptAttrib with a dropout ratio of 0.4 and 0.45 shows a decrease in F-score (0.376 and 0.391, respectively) compared to "PromptAttrib 0.35", indicating that too much dropout can degrade performance. However, the accuracy (A) metric is highest for "PromptAttrib 0.45" at 0.731, suggesting that while the F-score and average precision metrics are sensitive to dropout levels, accuracy might still benefit from higher dropout due to better generalization on certain aspects of the data.

8.5 Discussion

Prompt tuning can help us leverage the power of language models to solve different tasks. Considering attribute-level information can improve our ability to match two entities. By using fuzzy logic reasoning, we can better interpret the matching results. However, the prompt tuning method heavily relies on human-designed prompt templates. Finding different templates that are more suitable for various tasks is an important research direction and will be a key focus of our future research.

9 Dynamic Query Answering with Neural Symbolic Reasoning Over Incomplete KG

Conversational question answering over a knowledge graph involves engaging in natural, dialogue-like interactions with a system that uses a knowledge graph to provide accurate and relevant answers. By leveraging the rich symbolic triplet information in the knowledge graph, the system can deliver precise answers during the conversation. In this chapter, we discuss how to use neural network-based reinforcement learning methods to answer conversational questions with the help of symbolic knowledge from the knowledge graph.

9.1 Introduction

Knowledge graphs (KGs) are collections of nouns represented as nodes (representing real-world entities, events, and objects) and edges (denoting relationships between nodes). Knowledge graph question answering (KGQA) has long been a focus of study, with the goal of answering queries using information from a KG. However, traditional KGQA approaches often only consider single-shot questions, rather than the iterative nature of real-world conversation. Conversational question answering (ConvQA) addresses this gap by allowing users to interact with a QA system conversationally. ConvQA systems have had much success, as seen by Apple's Siri, Amazon's Alexa and OpenAI's ChatGPT.

Conversational question answering (ConvQA) involves a multiturn process consisting of users iteratively asking natural language questions, a system deciphering both the conversation context and underlying queries, and the system returning natural language answers. Some models will create rich, human-like responses [3, 12, 172], these methods are known as 'dialogue' conversation models. While for ConvQA over KGs, a corresponding entity in the KG is sufficient to answer the input question, we call it 'non-dialogue' conversation models. In this chapter, we focus on the non-dialogue ConvQA task as shown in Fig. 9.1.

L. Liu and H. Tong, *Neural Symbolic Knowledge Graph Reasoning*, Synthesis Lectures on Computer Science, https://doi.org/10.1007/978-3-032-15858-1_9

q_1: Who is the author that wrote the book Moby-Dick?

Reformulation1: Author of the book?
Reformulation2: Who wrote Moby Dick?

a^1: Herman Melville
q_2: When was he born?

Reformulation1: His birthdate is?
Reformulation2: When was Herman Melville born?

a^2: 1 August 1819
q_3: And where is he from originally?

Reformulation1: His place of birth?
Reformulation2: Where did he grow up?

a^3: Manhattan
q_4:How about his wife?

Reformulation1: Where is Herman Melville's wife from?
Reformulation2: Herman Melville's wife's place of birth?

a^4: Boston
q_5: Did they make a movie based on the book?
a^5: yes

Fig. 9.1 An example of conversational question answering

In general, a conversation is typically initiated with a well-formed question (i.e., q_1) followed by inexplicit follow-up questions (e.g., q_2–q_5). The initial question (q_1) often includes a central **topic entity** of interest ("Moby-Dick"), while the topic entities of follow-up questions (q_2–q_5) are not explicitly given. Additionally, the topic entity of the conversation may shift over time (e.g., inquiring about the birth time of Herman Melville in q_2).

To operate ConvQA over KG, different methods have previously been proposed. For instance, Magdalena et al. in [63] use named entity recognition (NER) methods to detect potential KG topic entities in the conversation and employ multi-agent reinforcement learning starting from these entities to find answers; the performance of this method is largely dependent on the quality of the detected entities. Philipp et al. in [29] propose finding a conversation-related subgraph and using heuristic-based methods to identify the answer within the subgraph. The subgraph is expanded as new questions are asked. Endri et al. in [62] use contrastive learning to make KG entity embeddings dissimilar from one another, enabling the model to separate correct answers from incorrect answers.

Despite the above achievements, inexplicit input data hinders a ConvQA system's ability to find correct answers [13, 63, 148]. Two common linguistic phenomena which undermine the semantic completeness of a query in the conversation are: **anaphora** and **ellipsis** [148]. **Anaphora** refers to the phenomenon of an expression that depends on an expression in the previous context. For example, in Example 1, the word "he" in q_2 refers to a^1. Meanwhile,

ellipsis refers to the phenomenon of the omission of expressions in the previous context. For example, the complete form q_4 should be "Where is Herman Melville's wife from?". To address this issue, several methods aim to learn a **reformulation** of the input query, *rewriting the original question* in a more meaningful way. Then, one can search for the answer using this new reformulation with existing techniques [13, 108, 148]. Despite many of the existing question rewriting models have shown potential to enhance ConvQA performance, as demonstrated by prior research [57], their generated reformulations fall short compared to human-generated reformulations [148].

In this chapter, we present **CORNNET**, a new reinforcement learning (RL) model for non-dialogue conversational question answering (ConvQA) with large language model (LLM) generated reformulations. First, we fine-tune existing LLMs, GPT2 [115] and Bart [73], to generated high quality reformulations, using human writing reformulations as the ground truth. Second, to further increase the convQA performance, we propose a teacher-student architecture to achieve near human-level performance.[1] Specifically, CORNNET (1) **directly trains** a teacher model with human writing reformulations in the training data, and (2) **indirectly trains** a student model with LLMs generated reformations to mimic the teacher model's output so that it can approach human-level performance. Note that the human writing reformulations only exist in the training and validation data. Lastly, to locate an answer, a RL model walks over the KG, sampling actions from a policy network to guide the direction of the walk and identify candidate answers. Our experiments demonstrate the effectiveness of CORNNET and its superiority over the state-of-the-art conversational question answering baselines.

9.2 Problem Definition

Table 9.1 gives the main notation used throughout this chapter. Uppercase letters are used for matrices, sets or constant value (e.g., C, T). Bold lowercase letters are for vectors or embedding (e.g., $\mathbf{r}_i$) and lowercase letters (e.g., s, a_i) for scalars or variables. A KG can be denoted as $\mathcal{G} = (\mathcal{V}, \mathcal{R}, \mathcal{L})$ where $\mathcal{V} = \{v_1, v_2, ..., v_n\}$ is the set of nodes/entities, $\mathcal{R} = \{r_1, r_2, ..., r_m\}$ is the set of relations and $\mathcal{L}$ is the list of triples. Each triple in the KG can be denoted as (h, r, t) where $h \in \mathcal{V}$ is the head (i.e., subject) of the triple, $t \in \mathcal{V}$ is the tail (i.e., object) of the triple and $r \in \mathcal{R}$ is the edge (i.e., relation, predicate) of the triple which connects the head h to the tail t. The embedding of a node or relation type is represented by bold lowercase letters, e.g., $\mathbf{e}_i$, $\mathbf{r}_i$. Each triple/edge (h, r, t) in the KG has a unique edge embedding which is denoted as $\mathbf{u}_r$.

[1] Human-level performance refers to the ability to find answers based on real human writing reformulations during testing.

Table 9.1 Notation and definitions

Symbols	Definition
$\mathcal{G} = (\mathcal{V}, \mathcal{R}, \mathcal{L})$	The knowledge graph
v_i	The ith entity/node in knowledge graph
r_i	The ith relation/edge in knowledge graph
$\mathbf{e}_{v_i}/\mathbf{e}_i$	The embedding of node v_i
$\mathbf{r}_i$	The embedding of relation r_i
$\mathbf{u}_{r_i}$	The unique embedding of edge r_i
C	The conversation
T	The number of turns in C
q_i	The ith question in C
$\text{Ref}^k_{q_i}$	The kth reformulation of question q_i
a^i	The answer of question q_i
w^t_i	The ith word in question q_t
v_{q_1}	The main topic entity of the conversation
$s \in S$	RL states
A_s	The set of possible actions at state s
$\mathbf{A}_s$	The corresponding matrix of A_s
$a_i \in A_s$	Actions a_i at state s
n_i	The entity in state s_i
R	Reward
θ	Parameters of policy network
π_θ	Policy parameterized by θ

Conversational question answering over a KG aims to iteratively answer multiple related questions from the users. Unlike dialog question answering which wants the chatbot to imitate the response of a human, ConvQA over KG only requires the model to return entities in the knowledge graph. We formally define the key terminologies used in this chapter as follows.

Conversation. A conversation C with T turns is made up of a sequence of questions $q_1, q_2, ..., q_T$ and their corresponding answers Ans = $\{a^1, a^2, ..., a^T\}$, such that $C = (q_1, a^1), (q_2, a^2), ..., (q_T, a^T)$. Example 1 in Introduction contains $T = 5$ turns. We assume that q_1 is well-formed, and all other q_t are inexplicit.

Question. Each question q_t is a sequence of words $q_t = (w^t_1, ..., w^t_{\Omega_t})$, where Ω_t is the number of words in q_t. We assume that each question can be mapped to a unique relation r_{q_t} in the KG and make no assumptions on the grammatical correctness of q_t.

Topic Entity. We assume that each q_t has a topic/central entity v_{q_t} which the user wants to ask about. We assume that the topic entity of q_1 is given in the training data, while the topic entities for other questions $q_2, ..., q_T$ are not given. For example, for the five questions in Example 1, their topic entities are `Moby Dick`, `Herman Melville`, `Herman Melville`, `Moby Dick` and `Moby Dick`, respectively. The topic entity of q_1 is presumed the main topic entity which is denoted as v_{q_1}.

Answer. Each answer a^t to question q_t is a (possibly multiple, single, or null-valued) set of entities in the KG. We assume that all the answer entities exist in the KG, except true or false questions.

Reformulation. A reformulation is a sentence which expresses the same information as the input question, but in a different way. We assume in the training data, each question has multiple reformulations.

Turn. Each question in C, including its reformulations and corresponding answers, constitutes a turn t_i. Each turn t_i contains a question q_i, the answer a^i and reformulations of q_i.

Based on the above, we formally define the problem of ConvQA over KG as:

Problem 9.1 Conversational Quesion Answering over Knowledge Graph

Given: (1) A knowledge graph G, (2) the training set of conversations where each question contains multiple human writing reformulations, (3) the test set of conversations where no question reformulation is provided;

Output: (1) The trained model, (2) the answer for each question in each conversation of the test set.

9.2.1 Preliminaries: Reinforcement Learning

The RL problem can usually be formulated as Markov Decision Processes (MDPs) [25, 26]. An MDP is defined by $M = (S, A, R, P, \gamma)$, where S is the state space and A is the action space. $R : S \times A \longrightarrow R$ is the reward function from the environment which maps a state-action pair to a scalar which denotes how much reward the agent can receive, and $P : S \times A \longrightarrow S$ is the transition function which defines the probability of transiting from a state-action pair to the next state. γ is a discount factor where $\gamma \in [0, 1]$. When modeling the KGQA problem as an RL task, the agent will learn a policy to find the correct answer in the KG, and make decisions guided by the policy function when it receives new queries.

9.3 Proposed Method

Due to anaphora and ellipsis, current ConvQA methods often rewrite input queries to generate more understandable reformulations. In this chapter, we follow this idea by fine-tuning two existing LLMs, GPT2 and Bart, to generate reformulations. Despite GPT2 and Bart are good reformulation generators, their performance still lags behind human-level performance. To further improve the performance, we propose a teacher-student architecture. The teacher model learns the question representation by using human writing reformulations, while the student model takes reformulations generated by LLMs as input, and tries to mimic

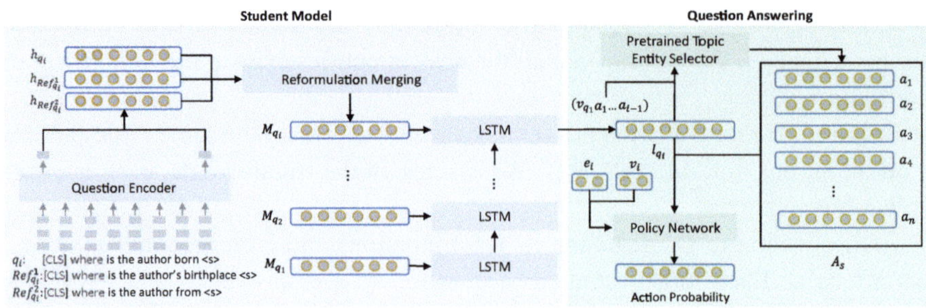

Fig. 9.2 Training Process of CORNNET. The light gray part denotes the architecture of the student model. The light green part shows the framework of RL-based question answering model. Both are trained end-to-end

the output of the teacher model, so that it can achieve the same performance as the teacher model despite using the LLMs generated reformulations.

In each iteration, our model uses the conversation history and the current query to identify the current topic entity, and an RL agent travels the KG starting from the topic entity to find the answer. This process is repeated for a number of turns until the conversation is completed. Figure 9.2 illustrates the framework of the proposed CORNNET. We will describe the details of each component in the following subsections.

9.3.1 Student Model: LLMs Reformulation Encoder

A—Context Encoder. Given a question $q_i = (w_1^i, w_2^i, ..., w_{\Omega_t}^i)$, we first add two indicator tokens ([CLS] and <s>) to the beginning and end of the question context to signify its boundary. Then, we pass the processed question context through a pre-trained BERT [37] to extract contextual embeddings for each token:

$$[\mathbf{h}_{CLS}, \mathbf{w}_1, \ldots, \mathbf{w}_{\Omega_t}, \mathbf{h}_\mathrm{s}] = \mathrm{BERT}([CLS], w_1, \ldots, w_{\Omega_t}, <s>) \tag{9.1}$$

where $\mathbf{h}_{CLS}$ is the embedding of the [CLS] token and $\mathbf{h}_\mathrm{s}$ is the embedding of the <s> token. The context question embedding is obtained from the transformation of $\mathbf{h}_{CLS}$ and $\mathbf{h}_\mathrm{s}$, where FFN is a feed forward neural network.

$$\mathbf{h}_{q_i} = \mathrm{FFN}(\mathbf{h}_{CLS} || \mathbf{h}_\mathrm{s}) \tag{9.2}$$

B—Context Fusion. During training, each input question has multiple corresponding reformulations generated. For each reformulation, we use the Context Encoder to obtain its context embedding. To merge the reformulation information, we stack the embeddings of the N

reformulations and the original question context embedding to create a sequence with N+1 embeddings. We treat this sequence as the embedding of a language sentence and pass it through a Transformer Encoder [149] to merge them together.

$$\mathbf{M_{q_i}} = \text{TRANSFORMER}([\mathbf{h}_{q_i}|\mathbf{h}_{Ref^1_{q_i}}|...|\mathbf{h}_{Ref^n_{q_i}}])[0] \tag{9.3}$$

where $\mathbf{M_{q_i}}$ is the query embedding after merging the reformulations.

C—Integrating Conversational History. Another problem in ConvQA is that the user's inputs are often ambiguous, hampering a system's ability to give accurate answers. This is illustrated in Example 1 q_3 "And where is he from originally?". It is impossible to identify the antecedent to the pronoun 'he' without any conversational history. Consequently, *conversational history is vital* to the success of CORNNET. We use an LSTM to encode all the conversational history which is given below.

$$\mathbf{l_{q_i}} = \text{LSTM}(\mathbf{M_{q_i}}) \tag{9.4}$$

the output of the LSTM $\mathbf{l_{q_i}}$ will be treated as the query embedding and be used by other components. Note that the reformulations used here are generated by LLMs.

9.3.2 Teacher Model: Human Writing Reformulation Encoder

Reformulations have been used by various methods to improve the performance of QA systems by creating more understandable queries. For instance, in [13], Christian et al. use a Seq2Seq-based reinforcement learning agent to transform input questions into machine-readable reformulations. In [148], Svitlana et al. propose a Transformer Decoder-based model for question rewriting. According to a study in [57], most question reformulation methods only improve the performance about 2–3%. Despite LLMs have exploded in popularity for all sorts of natural language tasks, the ConvQA performance based on LLMs generated reformulations is still upper bounded by human reformulations [57, 148].

To further improve the student model's performance, we propose a *teacher-student approach* where a teacher network is trained on human writing reformulations. The teacher network has the same network structure as the student model, but uses human writing reformulations as the input. An example is given in Fig. 9.3. During the training process, our goal is to make the output question embedding of the student model as close as possible to the output of the teacher model in the embedding space. The distance between the student's and teacher's output is measured using the L2 distance:

$$L = \sum_{q_i \in C} [d(\Upsilon_{q_i}, \mathbf{l}_{q_i})], \tag{9.5}$$

where Υ_{q_i} is the output of the teacher network for input question q_i, and l_{q_i} is the output of the student network for the same input with reformulations. By minimizing this distance,

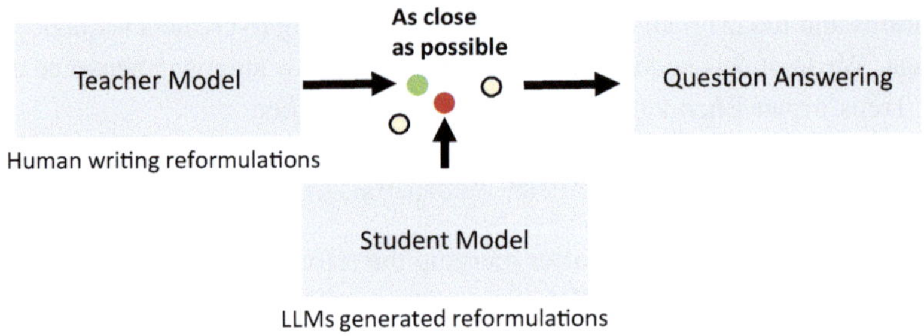

Fig. 9.3 Reformulation imitator

we can ensure that the student network is producing output that is similar to that of the teacher model, even when it only has access to the synthetic reformulations. Note that the teacher model is pretrained and fixed when we train the student model. During the testing phase, given a question, we first use LLMs to generate multiple reformulations for it, then the student model is used to encode the input question with LLMs generated reformulations. The performance of directly using the teacher model on the test data is slightly inferior to our model due to the different data distribution of human writing reformulations compared to the reformulations generated with LLMs.

9.3.3 Inferring the Topic Entity

During a conversation, the topic entity may change over time. To accurately answer questions, we determine the current topic entity based on the conversation history and the current question. We use a multi-layer perception (MLP) to determine whether the topic entity unchanged. If the classifier predicts that the topic entity of the current question is not the answer to the previous question, we set the topic entity to the main topic entity v_{q_1}.

The classifier consists of a feed forward neural network (FFN) with ReLU activation functions and a classification layer. The classification layer uses softmax on a 2D output to calculate the cross-entropy loss. Here we use the index format of PyTorch to show that the probability of $v_{q_i} = a^{i-1}$ is equal to the 2nd element of the 2D output.

$$\Pr(v_{q_i} = a^{i-1}) = \text{MLP\_classifier}(\text{FNN}(\mathbf{l}_{q_i}))[1] \tag{9.6}$$

A—Pretrain Topic Entity Selector. The goal of the Topic Entity Selector is to identify the correct topic entity within the conversation, which is an input to the RL model. In order to stabilize the training process for the RL model, we pre-train the parameters of the classifier using binary cross-entropy loss

$$\mathcal{L}_1 = -\left[y \log(\Pr(v_{q_i} = a^{i-1})) + \right. \quad (9.7)$$

$$\left. (1-y)\log(1 - \Pr(v_{q_i} = a^{i-1}))\right)] \quad (9.8)$$

9.3.4 Question Answering

After obtaining the topic entity, the next step is to find the correct entity to answer the user. We formulate this problem as a Markov decision process (MDP) which is defined by a 5-tuple (S, A, R, P, γ), where S is the state space, A is the action space, P is the state transition function and R denotes the reward function.

States. Intuitively, we want a state to encode the question, the current position of the agent in the KG, and the search history information. At the ith step, the state $s_i \in S$ is defined as a triple $s_t = (n_i, \mathbf{l}_q, \mathbf{g}_i)$, n_i is the current entity where the agent is at; $\mathbf{l}_q$ is the question embedding generated by the previous method; and $\mathbf{g}_i$ refers to the search history information. $(n_i, \mathbf{g}_i)$ can be viewed as state-dependent information while $(\mathbf{l}_q)$ is the global context shared by all states.

Actions. The set of possible actions A_s from a state $s_t = (n_t, \mathbf{l}_q, \mathbf{g}_t)$ consists of all outgoing edges of the vertex n_t in the KG. Formally, $A_s = \{(\mathbf{r}_i, \mathbf{u}_{r_i}, \mathbf{e}_{e'})|(n_t, r_i, e') \in G\}$. This means an agent at each state has the option to select which outgoing edge it wishes to take having knowledge of the label of the edge r_i and destination vertex e'. Note that different from most of the existing methods [63, 79] which only use $\mathbf{A}_s = (\mathbf{r}_i, \mathbf{e}_{e'})$, we also use the unique edge embedding $\mathbf{u}_{r_i}$. To allow the agent to have the option of ending a search, a self-loop edge is added to every entity. In addition, we also include the inverse relationship of a triple in the graph.

Transition. The transition function is defined as $\delta : S \times A \longrightarrow S$, which represents the probability distribution of the next states $\delta(s_{t+1}|s_t, a_t)$. In the current state s_t, the agent aims to choose proper actions a_t and then reach the next state $s_{t+1} = (n_{t+1}, \mathbf{l}_q, \mathbf{g}_{t+1})$. The n_t and g_t are updated, while the query and answer remains the same.

Rewards. The model will receive the reward of $R_b(s_t) = 1$ if the current location is the correct answer and 0 otherwise. We set $\gamma = 1$ during the experiments.

9.3.5 Policy Network

The search policy is parameterized using state information and global context, including the search history. Specifically, every entity and relation in $\mathcal{G}$ is assigned a dense vector embedding $\mathbf{e} \in \mathbb{R}^d$ and $\mathbf{r} \in \mathbb{R}^d$ respectively. The action $a_t = (\mathbf{r}_{r_i}, \mathbf{u}_{r_i}, \mathbf{e}_{e'}) \in A_t$ is represented as the concatenation of the relation embedding, the unique edge embedding and the end node embedding.

The search history $(n_1 = v_{q_i}, r_1, n_2, ..., n_t) \in H$ consists of the sequence of observations and actions taken up to step t, and can be encoded using an LSTM:

$$\mathbf{g}_0 = \text{LSTM}(0, [\mathbf{e}_{v_{q_i}} || \mathbf{l}_{q_i}]) \tag{9.9}$$

$$\mathbf{g}_t = \text{LSTM}(\mathbf{g}_{t-1}, \mathbf{a}_{t-1}), t > 0 \tag{9.10}$$

where $\mathbf{l}_{q_i}$ is the question embedding to form a start action with $\mathbf{e}_{v_{q_i}}$. The action space is encoded by stacking the embeddings of all actions in A_t: $\mathbf{A}_t \in \mathbb{R}^{|A_t| \times 3d}$. And the policy network π is defined as:

$$\pi_\theta(a_t|s_t) = \delta(\mathbf{A}_t \times \mathbf{W}_2\text{ReLU}(\mathbf{W}_1[\mathbf{n}_t||\mathbf{l}_{q_i}||\mathbf{g}_t])) \tag{9.11}$$

where δ is the softmax operator.

9.3.6 Knowledge-Based Soft Reward

Due to the weak supervision in ConvQA, the agent will receive a positive reward until it arrives at the target entity. Such delayed and sparse rewards significantly slow the convergence. To address the issue of weak supervision and sparsity of rewards in ConvQA, we assign a soft reward to entities other than the target answer to measure the similarity between them. This helps to speed up convergence and mitigate incompleteness in the KG. Specifically, the soft reward is used to measure the similarity between the current entity n_t identified by our model and the ground truth answer a^t. We use ComplEx [147] to learn the initial entity embedding and relation embedding for all nodes and edges in the knowledge graph. The probability that n_t is the correct answer is calculated by

$$Pr(n_t|\mathbf{l}_{q_i}, v_{q_i}, \mathcal{G}) = Re(< \mathbf{l}_{q_i}, \mathbf{e}_{n_t}, \bar{\mathbf{e}}_{v_{q_i}} >) \tag{9.12}$$

We propose the following soft reward calculating strategy

$$R(s_t) = R_b(s_t) + (1 - R_b(s_t))Pr(n_t|\mathbf{l}_{q_i}, v_{q_i}, \mathcal{G}) \tag{9.13}$$

Namely, if the destination n_t is a correct answer according to $\mathcal{G}$, the agent receives reward 1. Otherwise the agent receives a fact score weighted by a pretrained distribution: $Pr(n_t|\mathbf{l}_{q_i}, v_{q_i}, \mathcal{G})$.

9.3.7 Training

Given a set of conversations, we want to return the best possible answers a^*, maximizing a reward $a^* = argmax_a \sum_C \sum_T R(a^i|q_i)$. The reward is computed with respect to the question q_i while the answer is provided in the train dataset. The goal is to maximize the expected reward of the answer returned under the policy $E_{a_1,\ldots,a_T \sim \pi_\theta}[R(s_t)]$. Since it is difficult to compute the expectation, we use Monte Carlo sampling to obtain an unbiased estimate:

$$E_{a_1,\ldots,a_T \sim \pi_\theta}[R(s_t)] \approx \frac{1}{N}\sum_{i=1}^{N}\sum_{j=1}^{T} R(s_t)\pi_\theta(a_t|s_t) \tag{9.14}$$

In the experiment, we approximate the expected reward by running multiple rollouts for each training example. The number of rollouts is fixed, We set this number to 20. We use REINFORCE [155] to compute gradients for training.

$$\nabla_\theta\, E_{a_1,\ldots,a_T \sim \pi_\theta}[R(s_t)] = \sum_{i=1}^{T} \nabla_\theta \pi_\theta(a_t|s_t)R(s_t) \tag{9.15}$$

$$\approx \frac{1}{N}\sum_{i=1}^{N}\sum_{j=1}^{T} R(s_t)\ \nabla_\theta\ \log(\pi_\theta(a_t|s_t)) \tag{9.16}$$

Additionally, to encourage diversity in the paths sampled by the policy during training, we add an entropy regularization term to our cost function, as proposed in [63].

$$H_{\pi_\theta}(.,s) = -\sum_{a\in A_s} \pi_\theta(a|s)\log\pi_\theta(a|s) \tag{9.17}$$

$H_{\pi,\theta}$ is added to the update to ensure better exploration and prevent the agent from getting stuck in local optima. This final objective is:

$$E_{a_1,\ldots,a_T \sim \pi_\theta}[R(s_t)] + \lambda H_{\pi_\theta}(.,s) \tag{9.18}$$

After training, in the testing phase, given a query, we rank all the entities in the KG based on their probability of being the correct answer. We let the policy network keep top-k most likely paths according to beam search, and we rank them according to their possibilities. For all the other entities which are not in the top-k candidates, we use ComplEx [147] to rank them according to Eq. (9.12).

9.4 Experiments

In this section, we evaluate the performance of the proposed CornNet algorithm on several public datasets. Our aim is to answer the following questions:

- How accurate is the proposed CornNet algorithm for ConvQA?
- How efficient is the proposed CornNet algorithm?

9.4.1 Experimental Setting

We use two datasets in the experiments: ConvQuestions [29] and ConvRef [63]. ConvQuestions contains a total of 6,720 conversations, each with 5 turns. ConvRef contains a total of 6,720 conversations, each with 5 turns. All the conversations in ConvQuestions and ConvRef belong to one of the five domains: "Books", "Movies", "Soccer", "Music", and "TV Series". Each domain contains 1344 training conversations, 448 validation conversations and 448 test conversations. The details of these datasets can be found in Appendix.

Both ConvQuestions and ConvRef use Wikidata[2] as their background KG. However, the full Wikidata KG is extremely large, containing approximately 2 billion triples. Therefore, in the experiment, we sample a subset of triples from Wikidata. We first take the overlapped entities between the Wikidata and the QA datasets, and then we further obtain all the one-hop neighbours of these overlapped entities. The one-hop neighbors are retrieved from both the original data dump and also the entities' corresponding online Wikidata websites.

We compare the performance of our method, CORNNET, with four baselines: **Convex** [29]: this method detects answers to conversational utterances over KGs by first extracting a subgraph, then identifying answers in the subgraph. **Conqer** [63]: this is the current state-of-the-art baseline. It uses RL with reformulations to find answers in the KG. **OAT** [102]: this Transformer-based model takes a JSON-like structure as input to generate a Logical Form (LF) grammar that can model a wide range of queries on the graph. It finds answers by applying the LF. **Focal Entity** [71]: this is a novel graph-based model to find answer by graph neural network.

Two LLMs are used to generate reformulations for the input query, which are GPT2 [115] and Bart [73]. For each input question, we generate multiple reformulations and use attention mechanism to aggregate them inside the model. We adopt the following ranking metrics which are also employed by the previous baselines: (1) Precision at the top rank (P@1); (2) Mean Reciprocal Rank (MRR) is the average across the reciprocal of the rank at which the first context path was retrieved; (3) Hit ratio at k (H@k/Hit@k) is the fraction of times a correct answer was retrieved within the top-k positions.

9.4.2 Main Results

In this subsection, we test CORNNET on conversational question answering tasks and compare it with other baseline methods.

A—Overall performance on ConvQA datasets. Table 9.3 compares the results of CORNNET with baselines on the ConvQuestions and ConvRef datasets. As we can see, CORNNET outperforms the previous baselines in the H@5 and MRR metrics On ConvQA. For H@5, CORNNET performs 4.5% better than CONQUER and 20% better than CONVEX. In terms

[2] https://www.wikidata.org/wiki/Wikidata:Database_download.

of MRR, CONVEX has the lowest performance, which is 13.7% worse than CORNNET. CONQUER has the second highest performance, but it is also 1% lower than CORNNET. For OAT, because its source code is not available, we directly adopt its results from [102]. We can find that it has the highest P@1 compared to other methods. However, its MRR is 7.3% points lower than that of CORNNET. For Focal Entity, it has the second highest P@1 and the third highest MRR. On the ConvRef dataset, CORNNET also has similar performance. It achieves the highest Hit@5, which is 5% better than that of CONQUER. CORNNET also has the second highest P@1 and MRR compared with other baselines. Due to the unavailability of the OAT source code and the failure to run Focal Entity on the ConvRef dataset, we are unable to include their results in our analysis of the ConvRef dataset.

B—Performance on different domains. We further investigate the ranking performance of CORNNET across different domains for both benchmarks. Table 9.2 illustrates detailed ranking results for H@3, H@5 and H@8. As the results show, CORNNET outperforms other baselines on most domains in the ConvQuestions benchmark. On average, it achieves a 13.7% improvement in H@8 and 6.6% improvement in H@5 compared to the second highest baseline CONQUER. It performs slightly worse on H@3, with an average of 0.58% lower than CONQUER. The relatively poor performance on the Books domain is likely due to the presence of "yes/no" questions and queries regarding the plot, making it difficult for the topic entity selector to accurately determine the topic entity, posing a challenge for the model's predictions. Similar results are also observed on the ConvRef dataset.

9.4.3 Ablation Studies and Efficiency Results

In this subsection, we show the effectiveness of by ablation Studies.

A—The effectiveness of the reformulations. In this subsection, we demonstrate the effectiveness of using different reformulations in the model. Two large language models are used to generate reformulations: GPT2 and Bart. When not using reformulations, the output of the Question Encoder is treated as the input of the LSTM directly. Table 9.4 shows the experiment results. As we can see, using reformulations can indeed increase the ConvQA performance most of the time. The performance of using GPT2 reformulations is very similar to that of using Bart reformulations. Using human writing reformulations has the best performance.

We further test the effectiveness of the proposed teacher-student model, shown in Table 9.5. If we only use the reformulations generated by LLMs, the performance is about 3% lower than that of teacher-student model. If we train the model on human writing reformulations while tests on generated reformulations, the performance is about 1.3% lower than CORNNET.

C—Efficiency. Figure 9.4 shows the training time and test time of different methods on ConvRef dataset. As we can see, CORNNET has the shortest training and test time. While

Table 9.2 Performance on different domain datasets

Domain	Movies			TV series			Music			Soccer		
Metric	H@3	H@5	H@8	H@3	H@5	H@8	H@3	H@5	H@8	H@3	H@5	H@8
Dataset	ConvQA											
Conv	0.345	0.345	0.345	0.303	0.308	0.309	0.213	0.217	0.221	0.2019	0.2019	0.2019
Conquer	0.336	0.343	0.354	0.375	0.396	0.421	0.279	0.282	0.289	*0.325*	0.343	0.350
CornNet	*0.385*	*0.456*	*0.514*	*0.437*	*0.482*	*0.587*	*0.290*	*0.356*	*0.392*	0.288	*0.418*	*0.550*
Dataset	ConvRef											
Conv	0.345	0.345	0.345	0.303	0.308	0.308	0.213	0.217	0.221	0.202	0.202	0.202
Conquer	0.389	0.404	0.429	0.435	0.442	0.485	0.371	0.398	0.413	*0.371*	0.384	0.393
CornNet	*0.393*	*0.461*	*0.521*	*0.436*	*0.533*	*0.564*	*0.377*	*0.447*	*0.484*	0.353	*0.385*	*0.425*

Table 9.3 Performance on ConvQA and ConvRef

Dataset	ConvQA		ConvRef	
Model	Hit@5	MRR	Hit@5	MRR
CONVEX	0.219	0.200	0.257	0.241
CONQUER	0.372	0.327	0.427	*0.382*
OAT	–	0.260	–	–
Focal Entity	–	0.248	–	–
CORNNET	*0.417*	*0.337*	*0.477*	0.353

Table 9.4 The effectiveness of the reformulations

Model	P@1	Hit@3	Hit@5	Hit@8
No reformulation	0.231	0.373	0.445	0.474
GPT2 reformulation	0.212	0.367	0.451	0.504
Bart reformulation	0.221	0.376	0.441	0.494
Human reformulation	*0.257*	*0.395*	*0.463*	*0.518*

Table 9.5 The effectiveness of the teacher-student model

Model	P@1	Hit@3	Hit@5	Hit@8
GPT2 reformulation	0.212	0.367	0.451	0.504
CORNNET (GPT2)	*0.265*	*0.404*	*0.477*	*0.526*
Bart reformulation	0.221	0.376	0.441	0.494
CORNNET (Bart)	*0.244*	*0.413*	*0.491*	*0.523*
Train test different	0.237	0.389	0.472	0.507

Conv has the longest training and test time. Despite the short training time of CORNNET, it can still achieve better or comparable ConvQA performance compared to other baselines.

D—The effectiveness of the unique relation embedding. We also test the effectiveness of using a unique edge embedding in the action space with reinforcement learning. 4 reformulations are used in the experiments, and the results are shown in Table 9.6. In this table, "CORNNET -" denotes CORNNET without using a unique edge embedding. As we can see, using a unique edge embedding improves the question answering performance by an average of 2.2%.

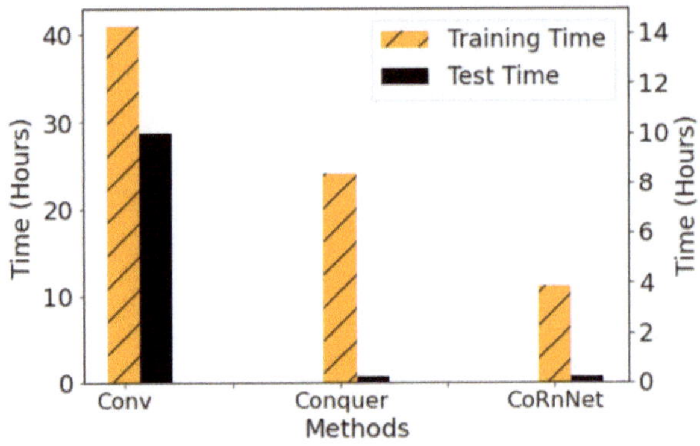

Fig. 9.4 CORNNET training and test time

Table 9.6 The effectiveness of the unique edge embedding

Model	P@1	Hit@3	Hit@5	Hit@8	MRR
CORNNET -	0.227	0.396	0.469	0.507	0.330
CORNNET	*0.265*	*0.404*	*0.477*	*0.526*	*0.353*

E—The effectiveness of the topic entity selector. The topic entity selector is pre-trained using the training data , and the teacher model is used to encode the input questions and reformulations during training. Note that a new teacher model will be retrained with the RL-based QA model after the topic entity selector is pre-trained. After pre-training, the results of the accuracy for the classifier is 83.2%. We find that the classifier has a relatively high accuracy.

9.5 Discussion

Reinforcement learning has commonly been used in conversational question answering over knowledge graphs. With the advent of large language models (LLMs), researchers are now exploring their use as reinforcement learning agents to find answers in knowledge graphs. The goal is to leverage the power of LLMs to better address this problem by enhancing their ability to understand and generate natural language responses. In the future, integrating human feedback during the reasoning process is a promising direction. Human feedback can provide valuable insights and corrections, which can improve the model's accuracy and reliability. By continuously refining the model with human input, we can develop more robust and trustworthy conversational agents.

Part III
Summary

Conclusion and Future Directions

10

In this chapter, we summarize our main research contributions and discuss the future research directions in knowledge graph reasoning.

10.1 Conclusion

In this book, we explore the problem of reasoning with knowledge graphs using neural symbolic techniques and develop several approaches to tackle the key challenges involved. Knowledge graphs represent real-world knowledge symbolically, and using neural networks to reason about them leverages the strengths of neural symbolic AI. Unlike neural networks, which often function as black boxes, knowledge graphs with their triplets, paths, and subgraphs provide more transparent explanations for reasoning results. However, achieving advanced neural symbolic AI still requires significant progress. Current neural-based models lack interpretability, and effectively integrating symbolic knowledge from knowledge graphs with neural networks remains an unsolved problem. While we've made strides, there is still much work ahead to reach the next level of neural symbolic AI. We summarize the key contributions of the book as follows.

Task 10.1—Accurate query reasoning over complete knowledge graph. We study the approximate attributed subgraph matching problem and develop an effective and scalable symbolic reasoning algorithm (G- FINDER). First, we propose a data structure called LOOKUP- TABLE (LTB) to efficiently index the candidates of the query graph. Second, we propose an effective algorithm to build and search LOOKUP- TABLE- GRAPH (LTBG) in order to find the *top-k* approximate subgraphs. Finally, we conduct extensive experiments on real world data, which demonstrate that the proposed G- FINDER (1) achieves an up to

L. Liu and H. Tong, *Neural Symbolic Knowledge Graph Reasoning*, Synthesis Lectures on Computer Science, https://doi.org/10.1007/978-3-032-15858-1_10

30% F1-Score improvement over the best baseline method, and (2) scales near-linearly to large data graphs with $1M+$ vertices and large query graphs with hundreds of nodes.

Task 10.2—Inconsistency detection through reasoning. We propose several symbolic reasoning methods to address comparative reasoning. At the core of these methods are key algorithms such as predicate-predicate similarity and symbolic semantic subgraph matching for extracting relevant knowledge segments. Additionally, we use symbolic influence function to identitfy important elements inside the subgraph. Experimental results show that the proposed methods (1) effectively detect semantic inconsistencies and (2) scale nearly linearly with the size of the knowledge graph.

Task 10.3—Accurate query reasoning over incomplete knowledge graph. We propose a neural symbolic reasoning method that simultaneously addresses multi-hop question answering (KGQA) and knowledge graph completion (KGC). By using a shared embedding space and answer scoring module for both tasks, they can learn from each other in a mutually beneficial way. Additionally, we developed a method that combines large language models (LLMs) with knowledge graph reasoning, allowing them to support each other. This approach mitigates the incompleteness of knowledge graphs and reduces the hallucination problem in LLMs.

Task 10.4—Reasoning for ambiguous query. We propose a neural symbolic model called PREFNET that uses pseudo relevance feedback to answer ambiguous queries over a knowledge graph. Guided by Bayes' rule, we break down the question-answering process into several components and use a neural network to model each one. To infer hidden queries, we use variational inference to update the complex model with latent variables. Additionally, we introduce a prompt-tuning based entity matching algorithm. By using both entity-level and attribute-level information, we can focus more on fine-grained details and achieve better accuracy.

Task 10.5—Reasoning for iterative query. We propose a neural symbolic model, CORNNET, that creatively combines question reformulation and reinforcement learning on a knowledge graph (KG) to achieve accurate multi-turn conversational question answering. CORNNET utilizes a teacher-student distillation approach and reinforcement learning to find answers from a KG. Experimental results demonstrate that CORNNET surpasses existing methods on various benchmark datasets on conversational question answering.

10.2 Future Directions

Despite our extensive research efforts, knowledge graph reasoning remains an active but not fully explored field, offering many potential opportunities. Advancing neural symbolic AI to the next level still requires significant work. In this section, we present several promising research directions.

Combining Large Language Models with Knowledge Graph Reasoning. Knowledge graphs provide accurate data and structured knowledge, while large language models (LLMs) excel in understanding language and solving various problems. By combining these two, we can create models that are more interpretable, integrate knowledge better, and make informed decisions. Our previous research showed that LLMs struggle with questions requiring multiple reasoning steps, whereas knowledge graph reasoning consistently yields accurate results for each step. Combining them can improve accuracy. Based on our previous observation, we mainly focus on doing research for the following three directions. Firstly, we delve into enhancing knowledge graph reasoning through Large Language Models, where large language models serve as a component in the reason process. Secondly, we study enhancing LLMs with knowledge graph reasoning, where knowledge graph reasoning is used to alleviate the hallucination problem. Lastly, we explore the integration of knowledge graph reasoning with LLMs in a mutually beneficial manner.

Additionally, existing methods often use LLMs as reinforcement learning agents, neglecting the role of human feedback. By involving humans in the loop, we can provide essential feedback and supervision, improving the system's adaptability and accuracy. This approach aims to develop advanced AI systems capable of understanding complex situations, making better decisions, and solving problems more effectively. This makes it a promising direction for the future.

The Application of Knowledge Graph Reasoning Across Various Domains. Knowledge graph reasoning plays a pivotal role in numerous high-impact applications, such as recommender systems, drug discovery, fact-checking and so on. However, current methods often focus on addressing a specific task within a single knowledge graph. In our upcoming research, we aspire to explore the sharing of information among diverse tasks and the transfer of knowledge between different graphs. Additionally, we aim to build foundation models or those large pretrained models for knowledge graphs, which can quickly adapt to different downstream tasks.

Knowledge Graph Empowered Neural-symbolic AI. Neural Symbolic AI combines the power of deep learning with symbolic reasoning, aiming to create more explainable, trustworthy and versatile AI systems that can harness the strengths of both symbolic and neural approaches. Since knowledge graphs can be viewed as discrete symbolic representations of knowledge, it is natural to integrate knowledge graphs with neural models to unleash the full potential of neural-symbolic reasoning. However, applying knowledge graphs to neural-symbolic AI is challenging. This is because symbolic systems rely on discrete symbolic representations, while neural systems employ continuous feature vector representations in the reasoning process. Consequently, a significant challenge lies in designing a unified framework that seamlessly integrates both symbolic and neural components. Building upon my completed works on multi-hop query reasoning [82, 86] and graph query reasoning [81, 85], I will investigate the design principles, algorithms, and systems to appropriately combine symbolic and neural systems to align with the requirements of different tasks and explore higher-level interactions between neural networks and symbolic reasoning. Addressing these challenges and advancing research in this direction will empower AI to an unprecedented level, fostering greater understanding and trust in AI systems.

References

1. Lamda: Language models for dialog applications (2022)
2. A. Ralph, I. Ceylan, T. Lukasiewicz, T. Salvatori, A box embedding model for knowledge base completion. Boxe
3. A. Acharya, S. Adhikari, Alexa conversations: an extensible data-driven approach for building task-oriented dialogue systems (2021)
4. P. Anchuri, M.J. Zaki, O. Barkol, S. Golan, M. Shamy, Approximate graph mining with label costs, in *Proceedings of the 19th ACM SIGKDD International Conference on Knowledge Discovery and Data Mining, KDD '13* (ACM, New York, NY, USA, 2013), pp. 518–526
5. N. Cao, B. Du, S. Zhang, H. Tong, First: Fast interactive attributed subgraph matching, in *Proceedings of the 23rd ACM SIGKDD* (2017)
6. D. Bahdanau, K. Cho, Y. Bengio, Neural machine translation by jointly learning to align and translate, in *International Conference on Learning Representations* (2016)
7. Y. Bai, X. Lv, J. Li, L. Hou, Answering complex logical queries on knowledge graphs via query computation tree optimization, in *Proceedings of the 40th International Conference on Machine Learning* (2023)
8. B. Maciej, B. Nils, K. Ales, G. Robert, G. Lukas, G. Joanna, L. Tomasz, P. Michal, N. Hubert, N. Piotr, T. Hoefler, Solving elaborate problems with large language models. Graph Thoughts (2023)
9. B. Bhattarai, H. Liu, H. Howie Huang, Ceci: compact embedding cluster index for scalable subgraph matching (2019)
10. F. Bi, L. Chang, X. Lin, L. Qin, W. Zhang, Efficient subgraph matching by postponing cartesian products, in *Proceedings of the 2016 International Conference on Management of Data, SIGMOD '16* (ACM, New York, NY, USA, 2016), pp. 1199–1214
11. A. Bordes, N. Usunier, Translating embeddings for modeling multi-relational data. Adv. Neural Inf. Process. Syst. **26** (2013)
12. T. Brown, B. Mann, N. Ryder, Subbiah, Language models are few-shot learners, in *Advances in Neural Information Processing Systems*, ed. by H. Larochelle, M. Ranzato, R. Hadsell, M.F. Balcan, H. Lin, vol. 33 (Curran Associates, Inc., 2020), pp. 1877–1901
13. C. Buck, J. Bulian, Ask the right questions: active question reformulation with reinforcement learning (2017)
14. C. Hanning, N. Yang, Z. Ali, Z. Zhuowen, Y. Sanggeon, W. Fei, K. Behnam, S. Narayan, L. Hugo, M. Imani, Algorithm-hardware codesign for hyperdimensional knowledge graph reasoning. Hdreason

L. Liu and H. Tong, *Neural Symbolic Knowledge Graph Reasoning*, Synthesis Lectures on Computer Science, https://doi.org/10.1007/978-3-032-15858-1

15. H. Chen, A. Zakeri, F. Wen, H. E. Barkam, M. Imani, Hypergraf: hyperdimensional graph-based reasoning acceleration on FPGA, in *2023 33rd International Conference on Field-Programmable Logic and Applications (FPL)* (2023), pp. 34–41
16. H. Chen, M. Liu, Y. Zhao, X. Yan, D. Yan, J. Cheng, G-miner: An efficient task-oriented graph mining system, in *Proceedings of the Thirteenth EuroSys Conference, EuroSys '18* (ACM, New York, NY, USA, 2018), pp. 32:1–32:12
17. J. Chen, T. Lan, N. Choi, Distributional-utility actor-critic for network slice performance guarantee, in *Proceedings of the Twenty-fourth International Symposium on Theory, Algorithmic Foundations, and Protocol Design for Mobile Networks and Mobile Computing* (2023), pp. 161–170
18. J. Chen, T. Lan, C. Joe-Wong, Rgmcomm: return gap minimization via discrete communications in multi-agent reinforcement learning, in *Proceedings of the AAAI Conference on Artificial Intelligence*, vol. 38 (2024), pp. 17327–17336
19. J. Chen, Y. Wang, T. Lan, Bringing fairness to actor-critic reinforcement learning for network utility optimization, in *IEEE INFOCOM 2021-IEEE Conference on Computer Communications* (IEEE, 2021), pp. 1–10
20. J. Chen, L. Zhang, J. Riem, G. Adam, N.D. Bastian, T. Lan, Explainable learning-based intrusion detection supported by memristors, in *2023 IEEE Conference on Artificial Intelligence (CAI)* (IEEE, 2023), pp. 195–196
21. J. Chen, L. Zhang, J. Riem, G. Adam, N.D. Bastian, T. Lan, Ride: real-time intrusion detection via explainable machine learning implemented in a memristor hardware architecture, in *2023 IEEE Conference on Dependable and Secure Computing (DSC)* (IEEE, 2023), pp. 1–8
22. J. Chen, H. Zhou, Y. Mei, G. Adam, N. D. Bastian, T. Lan, Real-time network intrusion detection via decision transformers. *arXiv preprint* arXiv:2312.07696 (2023)
23. X. Chen, Z. Hu, Y. Sun, Fuzzy logic based logical query answering on knowledge graphs, in *Proceedings of the AAAI Conference on Artificial Intelligence* (2022)
24. Y. Chen, H. Li, G. Qi, T. Wu, T. Wang, Outlining and filling: hierarchical query graph generation for answering complex questions over knowledge graph (2021)
25. Z. Chen, F. Silvestri, G. Tolomei, J. Wang, H. Zhu, H. Ahn, *Explain the explainer: interpreting model-agnostic counterfactual explanations of a deep reinforcement learning agent* (IEEE Trans. Artif, Intell, 2022)
26. Z. Chen, F. Silvestri, J. Wang, H. Zhu, H. Ahn, G. Tolomei, Relax: reinforcement learning agent explainer for arbitrary predictive models, in *Proceedings of the 31st ACM International Conference on Information & Knowledge Management* (2022), pp. 252–261
27. J. Cheng, Y. Ke, W. Ng, A. Lu, Fg-index: towards verification-free query processing on graph databases, in *Proceedings of the 2007 ACM SIGMOD International Conference on Management of Data, SIGMOD '07* (ACM, New York, NY, USA, 2007), pp. 857–872
28. N. Choudhary, C.K. Reddy, Complex logical reasoning over knowledge graphs using large language models (2024)
29. P. Christmann, R.S. Roy, A. Abujabal, J. Singh, G. Weikum, Look before you hop: conversational question answering over knowledge graphs using judicious context expansion, in *Proceedings of the 28th ACM International Conference on Information and Knowledge Management, CIKM '19* (Association for Computing Machinery, New York, NY, USA, 2019), pp. 729–738
30. G.L. Ciampaglia, P. Shiralkar, L.M. Rocha, Computational fact checking from knowledge networks (2015)
31. D.G. Corneil, C.C. Gotlieb, An efficient algorithm for graph isomorphism. J. ACM **17**(1), 51–64 (1970)

32. A. Creswell, M. Shanahan, I. Higgins, Selection-inference: exploiting large language models for interpretable logical reasoning, in *The Eleventh International Conference on Learning Representations* (2023)
33. C. Limeng, W. Suhang, D. Lee, Sentiment-aware multi-modal embedding for detecting fake news. Same (2019)
34. J.D. Moorman, Q. Chen, T.K. Tu, Z.M. Boyd, A. Bertozzi, Filtering methods for subgraph matching on multiplex networks (2018), pp. 3980–3985
35. D. Rajarshi, D. Shehzaad, M. Zaheer, Reasoning over paths in knowledge bases using reinforcement learning. Go for a walk and arrive at the answer (2017)
36. T. Dettmers, P. Minervini, P. Stenetorp, S. Riedel, Convolutional 2d knowledge graph embeddings (2017)
37. D. Jacob, C. Ming-Wei, L. Kenton, K. Toutanova, Pre-training of deep bidirectional transformers for language understanding. Bert (2019)
38. B. Dhingra, L. Li, X. Li, J. Gao, Y.-N. Chen, F. Ahmed, L. Deng, Towards end-to-end reinforcement learning of dialogue agents for information access, in *Proceedings of the 55th Annual Meeting of the Association for Computational Linguistics (Volume 1: Long Papers)* (Association for Computational Linguistics, Vancouver, Canada, 2017), pp. 484–495
39. B. Du, L. Liu, H. Tong, Sylvester tensor equation for multi-way association, in *Proceedings of the 27th ACM SIGKDD Conference on Knowledge Discovery and Data Mining, KDD '21* (Association for Computing Machinery, New York, NY, USA, 2021), pp. 311–321
40. E. Darren, T. Ha, C. Newman, B. Joshua, C. Alex, M. Apurva, T. Steven, J. Larson, A graph rag approach to query-focused summarization. From local to global (2024)
41. A. Elgohary, D. Peskov, J. Boyd-Graber, Can you unpack that? learning to rewrite questions-in-context, in *Proceedings of the 2019 Conference on Empirical Methods in Natural Language Processing and the 9th International Joint Conference on Natural Language Processing (EMNLP-IJCNLP)* (Association for Computational Linguistics, Hong Kong, China, 2019), pp. 5918–5924
42. A. Fader, L. Zettlemoyer, O. Etzioni, Open question answering over curated and extracted knowledge bases, in *Proceedings of the 20th ACM SIGKDD International Conference on Knowledge Discovery and Data Mining, KDD '14* (Association for Computing Machinery, New York, NY, USA, 2014), pp. 1156–1165
43. C. Faloutsos, K. McCurley, A. Tomkins, *Fast discovery of connection subgraphs, in KDD '04* (ACM, New York, NY, USA, 2004), pp.118–127
44. A. Fischer, C.Y. Suen, V. Frinken, K. Riesen, H. Bunke, Approximation of graph edit distance based on Hausdorff matching. Pattern Recogn. **48**(2), 331–343 (2015)
45. S. Freitas, N. Cao, Y. Xia, D.H.P. Chau, H. Tong, Local partition in rich graphs, in *BigData '19* (2018), pp. 1001–1008
46. G.W. Furnas, T.K. Landauer, L.M. Gomez, S.T. Dumais, The vocabulary problem in human-system communication (1987)
47. L.A. Gal'arraga, Amie: association rule mining under incomplete evidence in ontological knowledge bases, in *Proceedings of the 22nd International Conference on World Wide Web, WWW '13* (Association for Computing Machinery, New York, NY, USA, 2013)
48. T. Gao, X. Yao, D. Chen, SimCSE: simple contrastive learning of sentence embeddings, in *Empirical Methods in Natural Language Processing (EMNLP)* (2021)
49. C. Giovanni, S. Prashant, R. Luis, B. Johan, M. Filippo, F. Alessandro, Computational fact checking from knowledge networks. PLoS ONE **10**, 01 (2015)
50. A. Grover, J. Leskovec. node2vec: scalable feature learning for networks (2016)
51. D. Guo, D. Tang, N. Duan, M. Zhou, J. Yin, Dialog-to-action: conversational question answering over a large-scale knowledge base, in *Advances in Neural Information Processing Systems*, ed.

S. Bengio, H. Wallach, H. Larochelle, K. Grauman, N. Cesa-Bianchi, R. Garnett, vol. 31 (Curran Associates, Inc., 2018)

52. P. Hájek, *Metamathematics of Fuzzy Logic*, vol. 4 (Springer Science & Business Media, 2013)
53. S. He, K. Liu, G. Ji, J. Zhao, Learning to represent knowledge graphs with gaussian embedding, in *CIKM'15* (2015)
54. S. Hochreiter, J. Schmidhuber, Long short-term memory. Neural Comput. **9**(8), 1735–1780 (1997)
55. L. Hong, L. Zou, X. Lian, S. Philip, Subgraph matching with set similarity in a large graph database. IEEE Trans. Knowl. Data Eng. **27**(9), 2507–2521 (2015)
56. H. He, A.K. Singh, Closure-tree: an index structure for graph queries, in *22nd International Conference on Data Engineering (ICDE'06)* (2006), pp. 38–38
57. E. Ishii, Y. Xu, S. Cahyawijaya, B. Wilie, Can question rewriting help conversational question answering? (2022)
58. N. Jayaram, A. Khan, C. Li, X. Yan, R. Elmasri, Querying knowledge graphs by example entity tuples **27**(10), 2797–2811 (2015)
59. W. Jin, H. Yu, X. Tao, R. Yin, Improving embedded knowledge graph multi-hop question answering by introducing relational chain reasoning (2022)
60. P. Liang, K. Guu, J. Miller, Traversing knowledge graphs in vector space, in *Proceedings of the 2015 Conference on Empirical Methods in Natural Language Processing*
61. E. Kacupaj, J. Plepi, K. Singh, H. Thakkar, J. Lehmann, M. Maleshkova, Conversational question answering over knowledge graphs with transformer and graph attention networks, in *Proceedings of the 16th Conference of the European Chapter of the Association for Computational Linguistics: Main Volume* (Association for Computational Linguistics, 2021), pp. 850–862
62. E. Kacupaj, K. Singh, M. Maleshkova, J. Lehmann, Contrastive representation learning for conversational question answering over knowledge graphs (2022)
63. M. Kaiser, R.S. Roy, G. Weikum, Reinforcement learning from reformulations in conversational question answering over knowledge graphs, in *Proceedings of the 44th International ACM SIGIR Conference on Research and Development in Information Retrieval, SIGIR '21* (Association for Computing Machinery, New York, NY, USA, 2021), pp. 459–469
64. A. Khan, Y. Wu, C.C. Aggarwal, X. Yan, Nema: fast graph search with label similarity, in *Proceedings of the 39th International Conference on Very Large Data Bases, PVLDB'13* (VLDB Endowment, 2013), pp. 181–192
65. G. King, *Unifying Political Methodology: The Likelihood Theory of Statistical Inference* (University of Michigan Press, 1998)
66. D.P. Kingma, M. Welling, Auto-encoding variational bayes. arXiv (2013)
67. T. Kojima, S. Gu, M. Reid, Y. Matsuo, Y. Iwasawa, *Large language models are zero-shot reasoners* (Adv. Neural Inf. Process, Syst, 2022)
68. Y. Koren, S. North, C. Volinsky, *Measuring and extracting proximity in networks, in KDD '06* (ACM, New York, NY, USA, 2006), pp. 245–255
69. N. Lahn, S. Raghvendra, K. Zhang, A combinatorial algorithm for approximating the optimal transport in the parallel and MPC settings. Adv. Neural Inf. Process. Syst. **36** (2024)
70. Y. Lan, J. Jiang, Query graph generation for answering multi-hop complex questions from knowledge bases, in *Proceedings of the 58th Annual Meeting of the Association for Computational Linguistics (ACL)* (2020)
71. Y. Lan, J. Jiang, Modeling transitions of focal entities for conversational knowledge base question answering, in *Proceedings of the 59th Annual Meeting of the Association for Computational Linguistics and the 11th International Joint Conference on Natural Language Processing (Volume 1: Long Papers)* (Association for Computational Linguistics, 2021), pp. 3288–3297
72. L. Zhenzhong, C. Mingda, G. Sebastian, G. Kevin, S. Piyush, R. Soricut, A lite Bert for self-supervised learning of language representations. Albert (2020)

73. M. Lewis, Y. Liu, N. Goyal, Denoising sequence-to-sequence pre-training for natural language generation, translation, and comprehension. Bart (2019)
74. P. Lewis, E. Perez, A. Piktus, Retrieval-augmented generation for knowledge-intensive NLP tasks (2021)
75. Y. Li, J. Li, Y. Suhara, A. Doan, W.-C. Tan, Deep entity matching with pre-trained language models. *VLDB* (2020)
76. L. Ke, M. Lingyuan, L. Meng, L. Yue, W. Tu, W. Siwei, Z. Sihang, L. Xinwang, F. Sun, Static, dynamic, and multimodal. A survey of knowledge graph reasoning on graph types (2023)
77. D. Liben-Nowell, J. Kleinberg, The link prediction problem for social networks, in *CIKM '03*
78. P. Lin, Q. Song, W. Yinghui, Fact checking in knowledge graphs with ontological subgraph patterns. Data Sci. Eng. **3**(4), 341–358 (2018)
79. X.V. Lin, R. Socher, C. Xiong, Multi-hop knowledge graph reasoning with reward shaping, in *Proceedings of the 2018 Conference on Empirical Methods in Natural Language Processing, EMNLP* (Brussels, Belgium, 2018)
80. L. Liu, B. Du, H. Tong, Gfinder: approximate attributed subgraph matching, in *BigData '19* (2019)
81. L. Liu, B. Du, J. xu, H. Tong, G-finder: approximate attributed subgraph matching, in *2019 IEEE International Conference on Big Data (Big Data)* (2019), pp. 513–522
82. L. Liu, Y. Chen, M. Das, H. Yang, H. Tong, Knowledge graph question answering with ambiguous query, in *Proceedings of the ACM Web Conference 2023* (2023)
83. L. Liu, B. Du, Y.R. Fung, H. Ji, J. Xu, H. Tong, Kompare: a knowledge graph comparative reasoning system, in *Proceedings of the 27th ACM SIGKDD Conference on Knowledge Discovery and Data Mining, KDD '21* (Association for Computing Machinery, New York, NY, USA, 2021), pp. 3308–3318
84. L. Liu, B. Du, H. Ji, H. Tong, A knowledge graph reasoning prototype. *NeurIPS (demo track)* (2020)
85. L. Liu, B. Du, H. Ji, C. Zhai, H. Tong, Neural-answering logical queries on knowledge graphs, in *Proceedings of the 27th ACM SIGKDD Conference on Knowledge Discovery and Data Mining, KDD '21* (Association for Computing Machinery, New York, NY, USA, 2021), pp. 1087–1097
86. L. Liu, B. Du, J. Xu, Y. Xia, H. Tong, Joint knowledge graph completion and question answering, in *Proceedings of the 28th ACM SIGKDD Conference on Knowledge Discovery and Data Mining, KDD '22* (Association for Computing Machinery, New York, NY, USA, 2022), pp. 1098–1108
87. L. Liu, B. Hill, B. Du, F. Wang, H. Tong, Conversational question answering with reformulations over knowledge graph. *arXiv preprint* arXiv:2312.17269 (2023)
88. L. Liu, H. Ji, J. Xu, H. Tong, Comparative reasoning for knowledge graph fact checking, in *2022 IEEE International Conference on Big Data (Big Data)* (2022), pp. 2309–2312
89. L. Liu, H. Tong, Knowledge graph reasoning and its applications, in *Proceedings of the 29th ACM SIGKDD Conference on Knowledge Discovery and Data Mining* (2023), pp. 5813–5814
90. L. Liu, Z. Wang, J. Bai, Y. Song, H. Tong, New frontiers of knowledge graph reasoning: recent advances and future trends, in *Companion Proceedings of the ACM on Web Conference* (2024), pp. 1294–1297
91. L. Liu, Z. Wang, R. Qiu, Y. Ban, H. Tong, Logic query of thoughts: guiding large language models to answer complex logic queries with knowledge graphs. *arXiv preprint* arXiv:2404.04264 (2024)
92. L. Liu, R. Zhao, B. Du, Y.R. Fung, H. Ji, J. Xu, H. Tong, Knowledge graph comparative reasoning for fact checking: problem definition and algorithms. Data Eng. 19 (2022)
93. L. Liu, R. Zhao, B. Du, Y.R. Fung, H. Ji, J. Xu, H. Tong, Knowledge graph comparative reasoning for fact checking: problem definition and algorithms, in *2022 IEEE Data Engineering* (2022)

94. L. Yinhan, O. Myle, G. Naman, J. Du, J. Mandar, C. Danqi, L. Omer, L. Mike, Z. Luke, V. Stoyanov, A robustly optimized Bert pretraining approach. Roberta (2019)
95. J. Llad'os, E. Mart'ı, J.J. Villanueva, Symbol recognition by error-tolerant subgraph matching between region adjacency graphs. IEEE Trans. Pattern Anal. Mach. Intell. **23**(10), 1137–1143 (2001)
96. L. Linhao, L. Yuan-Fang, H. Gholamreza, S. Pan, Faithful and interpretable large language model reasoning. Reasoning on graphs (2024)
97. H. Ma, D. Zeng, Y. Liu, Learning individualized treatment rules with many treatments: a supervised clustering approach using adaptive fusion. Adv. Neural. Inf. Process. Syst. **35**, 15956–15969 (2022)
98. H. Ma, D. Zeng, Y. Liu, Learning optimal group-structured individualized treatment rules with many treatments. J. Mach. Learn. Res. **24**(102), 1–48 (2023)
99. J. Manyika, An overview of bard: an early experiment with generative AI (2023)
100. Y. Mao, S. Deb, S.B. Venkatakrishnan, S. Kannan, K. Srinivasan, Perigee: efficient peer-to-peer network design for blockchains, in *Proceedings of the 39th Symposium on Principles of Distributed Computing* (2020), pp. 428–437
101. Y. Mao, S.B. Venkatakrishnan, Less is more: understanding network bias in proof-of-work blockchains. Mathematics **11**(23), 4741 (2023)
102. P. Marion, P.K. Nowak, F. Piccinno, Structured context and high-coverage grammar for conversational question answering over knowledge graphs (2021)
103. A. Miller, A. Fisch, J. Dodge, A.-H. Karimi, A. Bordes, J. Weston, Key-value memory networks for directly reading documents (2016)
104. P. Minervini, E. Arakelyan, D. Daza, M. Cochez, Complex query answering with neural link predictors, in *International Conference on Learning Representations* (2021), pp. 1–14
105. T. Misu, K. Georgila, A. Leuski, D. Traum, Reinforcement learning of question-answering dialogue policies for virtual museum guides, in *Proceedings of the 13th Annual Meeting of the Special Interest Group on Discourse and Dialogue* (Association for Computational Linguistics, Seoul, South Korea, 2012), pp. 84–93
106. V. Nastase, R. Mihalcea, D.R. Radev, A survey of graphs in natural language processing. Natural Lang. Eng. **21**, 665–698 (2015)
107. M. Nickel, V. Tresp, H.-P. Kriegel, A three-way model for collective learning on multi-relational data, in *Proceedings of the 28th International Conference on International Conference on Machine Learning, ICML'11* (Omnipress, Madison, WI, USA, 2011), pp. 809–816
108. R. Nogueira, K. Cho, Task-oriented query reformulation with reinforcement learning (2017)
109. OpenAI. Gpt-4 technical report (2023)
110. M. Ou, P. Cui, J. Pei, Z. Zhang, W. Zhu, Asymmetric transitivity preserving graph embedding, in *Proceedings of the 22Nd ACM SIGKDD International Conference on Knowledge Discovery and Data Mining, KDD '16* (ACM, New York, NY, USA, 2016), pp. 1105–1114
111. P. Shirui, L. Linhao, W. Yufei, C. Chen, W. Jiapu, X. Wu, A roadmap, unifying large language models and knowledge graphs (2023)
112. F. Petroni, T. Rocktäschel, P. Lewis, A. Bakhtin, Y. Wu, A.H. Miller, S. Riedel, Language models as knowledge bases? (2019)
113. A. Phatak, S. Raghvendra, C. Tripathy, K. Zhang, Computing all optimal partial transports, in *International Conference on Learning Representations* (2023)
114. R. Pienta, A. Tamersoy, H. Tong, D.H. Chau, Mage: matching approximate patterns in richly-attributed graphs, in *2014 IEEE International Conference on Big Data (Big Data)* (2014), pp. 585–590
115. R. Alec, J. Wu, R. Child, D. Amodei, I. Sutskever, Language models are unsupervised multitask learners. David Luan (2019)

116. N. Reimers, I. Gurevych, Sentence-Bert: sentence embeddings using siamese bert-networks. *arXiv preprint* arXiv:1908.10084 (2019)
117. H. Ren, W. Hu, J. Leskovec, Query2box: reasoning over knowledge graphs in vector space using box embeddings, in *International Conference on Learning Representations* (2020)
118. H. Ren, J. Leskovec, Beta embeddings for multi-hop logical reasoning in knowledge graphs (2020)
119. X. Ren, J. Wang, Exploiting vertex relationships in speeding up subgraph isomorphism over large graphs. Proc. VLDB Endow. **8**(5), 617–628 (2015)
120. S. Roach, C. Ni, Canon: complex analytics of network of networks for modeling adversarial activities, in *2020 IEEE International Conference on Big Data (Big Data)* (2020), pp. 1634–1643
121. J. Robinson, D. Wingate, Leveraging large language models for multiple choice question answering, in *The Eleventh International Conference on Learning Representations* (2023)
122. K. Ruan, X. Di, Learning human driving behaviors with sequential causal imitation learning, in *Proceedings of the AAAI Conference on Artificial Intelligence*, vol. 36 (2022), pp. 4583–4592
123. K. Ruan, X. Di, Infostgcan: an information-maximizing spatial-temporal graph convolutional attention network for heterogeneous human trajectory prediction. Computers, **13**(6) (2024)
124. K. Ruan, J. Zhang, X. Di, E. Bareinboim, Causal imitation learning via inverse reinforcement learning, in *The Eleventh International Conference on Learning Representations* (2022)
125. S. Prashant, F. Alessandro, M. Filippo, C. Giovanni, Finding streams in knowledge graphs to support fact checking (2017), pp. 859–864
126. A. Saxena, A. Tripathi, P. Talukdar, Improving multi-hop question answering over knowledge graphs using knowledge base embeddings, in *Proceedings of the 58th Annual Meeting of the Association for Computational Linguistics* (2020), pp. 4498–4507
127. H. Shang, Y. Zhang, X. Lin, J.X. Yu, Taming verification hardness: an efficient algorithm for testing subgraph isomorphism. Proc. VLDB Endow. **1**(1), 364–375 (2008)
128. N. Shervashidze, P. Schweitzer, E.J. van Leeuwen, K. Mehlhorn, K.M. Borgwardt, Weisfeiler-lehman graph kernels. J. Mach. Learn. Res. **12**, 2539–2561 (2011)
129. B. Shi, T. Weninger, Discriminative predicate path mining for fact checking in knowledge graphs. Know.-Based Syst. **104**(C), 123–133 (2016)
130. B. Shi, T. Weninger, Discriminative predicate path mining for fact checking in knowledge graphs
131. S. Weijia, M. Sewon, M. Yasunaga, Retrieval-augmented black-box language models. Replug (2023)
132. Y. Su, H. Sun, B. Sadler, M. Srivatsa, Z. Yan, X. Yan, On generating characteristic-rich question sets for QA evaluation, in *Proceedings of the 2016 Conference on Empirical Methods in Natural Language Processing* (Association for Computational Linguistics, 2016)
133. Y. Su, S. Yang, H. Sun, M. Srivatsa, S. Kase, M. Vanni, X. Yan, *Exploiting relevance feedback in knowledge graph search, in KDD '15* (New York, NY, USA, Association for Computing Machinery, 2015), pp. 1135–1144
134. F.M. Suchanek, G. Kasneci, G. Weikum, Yago: a core of semantic knowledge, in *WWW'07* (Association for Computing Machinery, 2007)
135. H. Sun, T. Bedrax-Weiss, W.W. Cohen, Pullnet: open domain question answering with iterative retrieval on knowledge bases and text (2019)
136. H. Sun, B. Dhingra, M. Zaheer, K. Mazaitis, R. Salakhutdinov, W.W. Cohen, Open domain question answering using early fusion of knowledge bases and text (2018)
137. S. Jiashuo, C. Xu, T. Lumingyuan, W. Saizhuo, L. Chen, G. Yeyun, M. Ni Lionel, S. Heung-Yeung, J. Guo, Deep and responsible reasoning of large language model on knowledge graph. Think-on-graph (2024)

138. Z. Sun, Z. Deng, J. Nie, J. Tang, Rotate: knowledge graph embedding by relational rotation in complex space. *CoRR*
139. I. Sutskever, J. Martens, G. Hinton, Generating text with recurrent neural networks, in *Proceedings of the 28th International Conference on International Conference on Machine Learning, ICML'11* (Omnipress, 2011)
140. A. Tchechmedjiev, P. Fafalios, K. Boland, M. Gasquet, M. Zloch, B. Zapilko, S. Dietze, K. Todorov, Claimskg: a knowledge graph of fact-checked claims, in *International Semantic Web Conference* (Springer, 2019), pp. 309–324
141. Y. Tian, R.C. Mceachin, C. Santos, D.J. States, J.M. Patel, Saga: a subgraph matching tool for biological graphs. Bioinformatics **23**(2), 232–239 (2007)
142. Y. Tian, J.M. Patel, Tale: a tool for approximate large graph matching, in *2008 IEEE 24th International Conference on Data Engineering* (2008), pp. 963–972
143. H. Tong, C. Faloutsos, *Center-piece subgraphs: problem definition and fast solutions, in KDD '06* (ACM, New York, NY, USA, 2006), pp.404–413
144. H. Tong, C. Faloutsos, C. Faloutsos, B. Gallagher, T. Eliassi-Rad, Fast best-effort pattern matching in large attributed graphs, in *Proceedings of the 13th ACM SIGKDD International Conference on Knowledge Discovery and Data Mining, KDD '07* (ACM, New York, NY, USA, 2007), pp. 737–746
145. H. Touvron, T. Lavril, Llama: open and efficient foundation language models (2023)
146. H. Touvron, L. Martin, Llama 2: open foundation and fine-tuned chat models (2023)
147. T. Trouillon, J. Welbl, S. Riedel, E. Gaussier, G. Bouchard, Complex embeddings for simple link prediction, in *Proceedings of the 33rd International Conference on International Conference on Machine Learning—Volume 48, ICML'16* (2016). JMLR.org
148. S. Vakulenko, S. Longpre, Z. Tu, R. Anantha, Question rewriting for conversational question answering, in *Proceedings of the 14th ACM International Conference on Web Search and Data Mining, WSDM '21* (Association for Computing Machinery)
149. V. Ashish, S. Noam, P. Niki, U. Jakob, L. Jones, A.N. Lukasz Kaiser, I. Polosukhin, Attention is all you need. Gomez (2017)
150. H. Wang, H. Ren, J. Leskovec, Relational message passing for knowledge graph completion, in *Proceedings of the 27th ACM SIGKDD Conference on Knowledge Discovery and Data Mining, KDD '21* (2021)
151. P. Wang, X. Zeng, L. Chen, F. Ye, Y. Mao, J. Zhu, Y. Gao, Promptem: prompt-tuning for low-resource generalized entity matching. VLDB (2022)
152. Z. Wang, K. Zhao, H. Wang, X. Meng, J. Wen, Query understanding through knowledge-based conceptualization, in *IJCAI'15* (AAAI Press, 2015), pp. 3264–3270
153. Z. Wang, H. Yin, Y. Song, Benchmarking the combinatorial generalizability of complex query answering on knowledge graphs, in *NeurIPS Datasets and Benchmarks Track* (2021)
154. J. Wei, X. Wang, D. Schuurmans, M. Bosma, F. Xia, E. Chi, Q.V. Le, D. Zhou et al., *Chain-of-thought prompting elicits reasoning in large language models* (Adv. Neural Inf. Process, Syst, 2022)
155. R.J. Williams, *Simple statistical gradient-following algorithms for connectionist reinforcement learning* (Mach, Learn, 1992)
156. T. Xie, K. Dai, K. Wang, R. Li, L. Zhao, Deepmatcher: a deep transformer-based network for robust and accurate local feature matching. Expert Syst. Appl. **237**, 121361 (2024)
157. X. Wenhan, H. Thien, W.Y. Wang, A reinforcement learning method for knowledge graph reasoning. Deeppath (2017)
158. X. Yan, P.S. Yu, J. Han, Graph indexing: a frequent structure-based approach, in *Proceedings of the 2004 ACM SIGMOD International Conference on Management of Data, SIGMOD '04* (ACM, New York, NY, USA, 2004), pp. 335–346

159. Y. Yan, H. Yongyi, Q. Zhou, L. Liu, Z. Zeng, Y. Chen, M. Pan, H. Chen, M. Das, H. Tong, Pacer: network embedding from positional to structural, in *Proceedings of the ACM on Web Conference* (2024), pp. 2485–2496
160. Y. Yan, B. Jing, L. Liu, R. Wang, J. Li, T. Abdelzaher, H. Tong, Reconciling competing sampling strategies of network embedding. Adv. Neural Inf. Process. Syst. **36** (2024)
161. Y. Yan, L. Liu, Y. Ban, B. Jing, H. Tong, Dynamic knowledge graph alignment. Proc. AAI Conf. Artif. Intell. **35**(5), 4564–4572 (2021)
162. B. Yang, W.T. Yih, X. He, J. Gao, L. Deng, Embedding entities and relations for learning and inference in knowledge bases (2015)
163. S. Yang, W. Yinghui, H. Sun, X. Yan, Schemaless and structureless graph querying. Proc. VLDB Endow. **7**(7), 565–576 (2014)
164. S. Yao, D. Yu, J. Zhao, I. Shafran, T.L. Griffiths, Y. Cao, K. Narasimhan, *Tree of thoughts: deliberate problem solving with large language models* (Adv. Neural Inf. Process, Syst, 2023)
165. X. Ye, S. Yavuz, K. Hashimoto, Y. Zhou, C. Xiong, Rng-kbqa: generation augmented iterative ranking for knowledge base question answering, in *Proceedings of the Annual Meeting of the Association for Computational Linguistics (ACL)* (2021)
166. H. Yin, Z. Wang, Y. Song, Rethinking complex queries on knowledge graphs with neural link predictors, in *The Twelfth International Conference on Learning Representations* (2024)
167. Q. Zhang, X. Weng, G. Zhou, Y. Zhang, J.X. Huang, Arl: an adaptive reinforcement learning framework for complex question answering over knowledge base. Inf. Process. Manage. **59**(3), 102933 (2022)
168. S. Zhang, J. Yang, W. Jin, Sapper: subgraph indexing and approximate matching in large graphs. Proc. LDB Endow. **3**(1–2), 1185–1194 (2010)
169. P. Zhao, J. Han, On graph query optimization in large networks. Proc. VLDB Endow. **3**(1–2), 340–351 (2010)
170. P. Zhao, J.X. Yu, P.S. Yu, Graph indexing: tree + delta > graph, in *Proceedings of the 33rd International Conference on Very Large Data Bases, VLDB '07* (2007), pp. 938–949
171. D. Zhou, N. Scharli, L. Hou, J. Wei, N. Scales, X. Wang, D. Schuurmans, C. Cui, Olivier Bousquet. Least-to-most prompting enables complex reasoning in large language models, in *The Eleventh International Conference on Learning Representations*, ed. by Q.V. Le, H. Chi (2023)
172. L. Zhou, J. Gao, D. Li, H.-Y. Shum, The design and implementation of XiaoIce, an empathetic social chatbot. Computat. Linguist. **46**(1), 53–93 (2020)
173. Q. Zhou, L. Li, N. Cao, L. Ying, H. Tong, Adversarial attacks on multi-network mining: problem definition and fast solutions, in *ICDM '19* (2019)
174. J. Zhuang, M. Al Hasan, Defending graph convolutional networks against dynamic graph perturbations via Bayesian self-supervision, in *Proceedings of the AAAI Conference on Artificial Intelligence*, vol. 36, no. 4 (2022), pp. 4405–4413
175. J. Zhuang, M. Al Hasan, How does bayesian noisy self-supervision defend graph convolutional networks? Neural Process. Lett. 1–22 (2022)

MIX
Papier aus verantwortungsvollen Quellen
Paper from responsible sources
FSC® C105338

If you have any concerns about our products,
you can contact us on
ProductSafety@springernature.com

In case Publisher is established outside the EU,
the EU authorized representative is:
Springer Nature Customer Service Center GmbH
Europaplatz 3, 69115 Heidelberg, Germany

Printed by Libri Plureos GmbH
in Hamburg, Germany